PLOUGHS, CHAFF CUTTERS AND STEAM ENGINES

LINCOLNSHIRE'S AGRICULTURAL IMPLEMENT MAKERS

Edited by Ken Redmore

Published by
THE SOCIETY FOR LINCOLNSHIRE HISTORY AND ARCHAEOLOGY
2007

PLOUGHS, CHAFF CUTTERS AND STEAM ENGINES

First published by the Society for Lincolnshire History and Archaeology 2007
© The Society for Lincolnshire History and Archaeology

ISBN 0 903582 30 9
978 0 903582 30 8

British Library Cataloguing in Publication Data
A CIP catalogue record for this book is available from the British Library

All rights reserved

No part of this publication may be reproduced or transmitted in any form or by any means,
electronic or mechanical, including photocopying, recording or any information storage and retrieval system,
without permission in writing from the publisher.

Cover photograph:
Tuxford steam threshing set of 1871 in use on the Thorold estate at Syston, near Grantham, up to 1914 *(Fig 170, page 133)*
The Museum of Lincolnshire Life, by courtesy of Lincolnshire County Council

Designed by Nigel Kingston

Printed in the United Kingdom by F W Cupit (Printers) Limited
Horncastle, Lincolnshire

PLOUGHS, CHAFF CUTTERS AND STEAM ENGINES

CONTENTS

Acknowledgements	4
Notes on Contributors	5
Introduction Ken Redmore	6
1: Blackstone & Co of Stamford: Engineers and Machine Makers *Michael Key*	8
2: John Cooke & Sons of Lincoln: Prize Plough Maker *Hugh Cooke*	24
3: James Coultas of Grantham: Prize Drill Maker *Catherine Wilson*	36
4: Richard Duckering of Lincoln: Ironfounder *Mark Duckering*	46
5: J & B Edlington of Gainsborough: Machine Makers *Susan Edlington, Tony Wall & Terry Maidens*	56
6: T & J Fletcher of Winterton: Agricultural Implement Makers *Charles Parker*	66
7: James Hart & Son of Brigg: Implement and Machine Makers *Chris Page*	76
8: C L Hett of Brigg: Hydraulic Engineer *Chris Page*	85
9: William Howden and Son of Boston: Steam Engine Pioneer *Neil Wright*	95
10: The Malleable, North Hykeham: Castings for Agriculture and the Motor Industry *Norman Tate*	103
11: Peacock & Binnington of Brigg: Agents for Tractors and Agricultural Machinery *Philip Brown*	112
12: John H Rundle of New Bolingbroke: Ironfounder and Fairground Ride Maker *Alan Rundle*	121
13: Tuxford and Sons of Boston: Steam Engine Builders *Neil Wright*	131
Lincolnshire Implement Makers: A Selected List	144
Glossary of Terms	150
Selected Booklist	153
Index	154

ACKNOWLEDGEMENTS

The East Midlands Industrial Archaeology Conference has contributed to the funding of this publication.

The following have generously provided information or advice:

Introduction: Catherine Wilson

Chapter 3 (Coultas): Lurleen Slaney, Ann King, Mary Whittington, Alan Rundle, Hilary Healey, Dr Roy Brigden, The Museum of Lincolnshire Life, Grantham Museum;

Chapter 4 (Duckering): Lincoln Central Library, Alex Wilcockson, Anna Richards, Clare Lee;

Chapter 6 (Fletcher): John T Fletcher

Chapter 8 (Hett): Jon Sass, Nick Lyons, Barry M Barton, Ms A Ruklmann (Institute of Civil Engineers), A Crocker;

Chapter 10 (The Malleable): Dave Grundy, Nev Gray;

Chapter 11 (Peacock & Binnington): Michael Peacock and employees of Peacock and Binnington, the late Edward Dodd;

Glossary: Chris Page, Catherine Wilson.

The following individuals and organisations have given permission to publish photographs and other illustrations in this book. (An unidentified source indicates that the illustration is either from the contributor's own collection or the Society for Lincolnshire History and Archaeology.)

Chapter 1 (Blackstone): Figs 1-7, 18-20; The Blackstone Collection; 8-17, 21 Lincolnshire County Council: Stamford Museum;

Chapter 2 (Cooke): 24 Robin Wheeldon; 25, 26 Barry Parr; 33, 35, 36 Robin Cooke;

Chapter 3 (Coultas): 39, 40, 45, 48, 54 Alan Rundle; 38, 42 Lurleen Slaney; 41 Ann King; 43 Lincolnshire Archives; 46, 47, 49-51 Museum of English Rural Life, Reading;

Chapter 4 (Duckering): 56, 57, 59-63 Lincolnshire County Council: Lincoln Central Library;

Chapter 5 (Edlington): 65-79 The Edlington Archive

Chapter 6 (Fletcher): 80-90, 92, 93 John T Fletcher; 91 Hull & Grimsby Newspapers

Chapter 7 (Hart): 94 Lincolnshire Archives; 95, 98 Lincolnshire Archives and Hett Johnson Whiting Solicitors; 96 Lincolnshire Archives and the Eighth Earl of Yarborough: 97 Lincolnshire County Council: Lincoln Central Library;

Chapter 8 (Hett): 103 Lincolnshire Agricultural Society; 104 Chris Lester; 105, 108 Museum of Rural English Life, Reading; 106, 107 Wheeler: Drainage of the Fens (1888); 110-112 North East Lincolnshire Archives Office;

Chapter 9 (Howden): 119 Alan Rundle; 122 Lincolnshire County Council: Lincoln Central Library;

Chapter 10 (Malleable): 124-128 Cyril Lyall; 130-132, 134-137 George Tokarski (David Flintham);

Chapter 11 (Peacock & Binnington): 138-150 Michael Peacock and Peacock & Binnington;

Chapter 12 (Rundle): 160 Ken Redmore;

Chapter 13 (Tuxford): 168, 175-177 The Museum of Lincolnshire Life, by courtesy of Lincolnshire County Council; 184 Ken Redmore

Index: 186-190 Museum of English Rural Life, Reading

CONTRIBUTORS

Philip Brown lives in Brigg and has worked for Peacock and Binnington for almost twenty years. He is a member of the Cromwell Association and has a BA (Hons) Degree in English and History from the University of Lincoln.

Hugh Cooke has worked in optical instrumentation, microwave electronics and telecommunications companies as a development engineer and in worldwide industrial sales and management. He is a great-grandson of John Cooke, the founder of the agricultural engineering company.

Mark Duckering, a born and bred Lincoln man, works as a finisher with a local printing firm. He is great-great-great-nephew of Richard Duckering, the founder of the Lincoln firm of ironfounders.

Susan Edlington is the company secretary and a director of the family business. She is deputy chairman of The Gainsborough Heritage Group and is author of several books relating to the history of Gainsborough, its people and its businesses. Her late husband Brian was great-great-nephew of J B and T Edlington, founders of the company.

Michael Key, now retired, spent the last fifteen years of his working life at Stamford Museum. His lifetime interest in industrial history has resulted in his writing a number of books and articles on the subject. He is publisher of *The Blackstone Collection*, a quarterly journal devoted to the history of Blackstone & Co Ltd.

Terry Maidens is a retired pharmacist whose interests include photography and industrial archaeology. He has co-operated with Susan Edlington on several projects to create digital archives. Terry is a former member of the SLHA Industrial Archaeology team.

Chris Page trained in museum studies, engineering and agriculture and has worked in various rural museums throughout the country. He recently returned to Lincolnshire and is working on a number of research projects in a freelance capacity.

Charles Parker was born in Winterton and spent the first part of his working life as an accountant at local steelworks and chemical producers. An SLHA member, he initially came into contact with the Fletcher family through the Royal Observer Corps in the early 1960s.

Ken Redmore had a career in schools, colleges and education administration before retirement. He is a member of the Industrial Archaeology Team at SLHA and has particular interest in farming technology and buildings.

Alan Rundle is grandson of John H Rundle and has been immersed in the work of the firm from his earliest days. He attends many of the UK steam rallies each year with the family's showman's engine and is widely respected for his expertise in steam engine technology.

Norman Tate served an engineering apprenticeship with Ruston & Hornsby and worked as an engineer officer with P&O for 5 years. He was maintenance engineer and later production line foreman at Leys Malleable in North Hykeham between 1963 and 1982. Norman Tate died in November 2006.

Tony Wall is a chartered insurance broker with a lifetime interest in all aspects of history. He has been a member of the SLHA Industrial Archaeology Team for nearly thirty-five years.

Catherine Wilson is a retired museum director. She has a long-standing interest in 'the hardware' of rural history and particularly in the products of Lincolnshire's agricultural engineering companies. She is a previous Chairman of the Industrial Archaeology Team and is currently President of SLHA.

Neil Wright has been a Committee Officer with Lincolnshire County Council and its predecessors for over 40 years. He has been a member of the Industrial Archaeology Team of SLHA since the 1960s and is the author of *Lincolnshire Towns and Industry 1700-1914* published by SLHA in 1982.

INTRODUCTION

Ken Redmore

During the nineteenth century the farming landscape in Lincolnshire was transformed. In the first few decades the process of enclosure was completed, fenland and marsh were effectively drained and cultivated and the wide swathes of upland, previously devoted to sheep walks and rabbit warrens, were brought under the plough. As the century progressed crop yields increased considerably, following improved field drainage, seed selection and the application of fertilisers. There were big improvements too in livestock quality through selective breeding and the widespread use of feedstuffs such as oil-seed cake from outside the farm. As a result, especially during the forty years from 1835 to 1875, most farmers and landowners were able to make reasonable profits. At the same time, the value of using cost effective and easily operated machinery for many of the routine tasks in the field, farmyard and barn was recognised. Such machinery increased the efficiency of many processes, reduced labour costs and as a result enhanced profits. Thus there was strong incentive for the development and improvement of implements for the many tasks of the farming year, from ploughing the soil to gathering the crop. During the nineteenth century the opportunities for machine and implement makers were considerable and, as was to be expected, businesses soon appeared in every sizeable community.

What range of products would the typical local implement and machine maker offer? Indeed which implements would the farmer with a mixed holding of a reasonable size be expected to use at this time? First and foremost was the plough, which was essential for turning over the soil between one growing season and the next. Numerous models were developed to suit differing soil conditions and local preferences; new designs and new materials evolved, especially for the all-important coulters, ploughshares and mouldboards. Then, after the roughly ploughed soil had weathered, the farmer had the task of producing a fine tilth suitable for sowing seed, and for this he needed a sequence of clod-crushers, rollers and harrows. What he actually chose to use - like so many things in farming - depended on local conditions, tradition and personal preference.

The traditional skill of sowing seed by hand had been largely superseded by the 1850s when efficient mechanical seed drills for corn (with manure dropped alongside if required) and other crops were marketed on a large scale. Sowing crops like turnips in regular, widely spaced rows opened up the possibility of weeding the growing plants with a horse-drawn hoe - an enormous advantage over previous cultivation methods.

In high summer the farmer turned his attention to haymaking, and for this activity he relied on a large labour force using hand tools. Haymaking was slow to be mechanised, with mowing machines only coming into widespread use in the last quarter of the nineteenth century. Horse-drawn tedders and rakes for turning over and gathering up the drying hay in the field were more common a little earlier.

The crucial period of the Lincolnshire farming year was the corn harvest, when again the farmer would marshal all the labour available to him. The relatively expensive mechanical reaper for cutting the corn first appeared on the larger farms in the 1850s and the prototype binder/reaper soon after, though its labour-saving potential was not fully realised until the introduction of the ingenious knot-tying mechanism, which bound the sheaves automatically with twine, in the 1880s. All the other tasks of getting the sheaves of corn safely from the field into barn or stack were achieved by manpower with the aid of simple hand tools and the ubiquitous horse and wagon.

A range of small compact machines, usually housed in the farm buildings and collectively known as barn machinery, was developed to crush oil-seed cake, to slice or pulverise turnips, to cut chaff or straw and to grind or crush grain and pulses. They all supported the regular task of feeding the animals in and around the farmstead, especially in the winter months.

The other frequent autumn or winter occupation was threshing corn, and for this a small proportion of barns were permanently fitted with threshing machines. However, in Lincolnshire the much more common arrangement was the portable threshing 'drum' brought into the stackyard and powered by horse-gear or steam engine, probably as part of a hired threshing set.
On smaller farms corn was still threshed by hand and an essential item in the barn would be a winnowing machine for cleaning the grain afterwards. Finally, every farmstead possessed a number of horse-drawn carts, wagons and 'moffreys' for the year-round tasks of carrying materials to and from the fields and also between farm and market.

This then was the range of implements and machines - together with a considerable number of hand tools - that most nineteenth-century Lincolnshire farmers expected to have. Very quickly, small manufacturing firms began to appear in large numbers across the county in response to this need and products were widely advertised and marketed. Simple cast-iron pieces and easily fabricated items were offered by most firms. More complex or technically sophisticated implements like drills and ploughs appeared less frequently; they were the specialisms of relatively few businesses and commanded a premium.

What can be said about the first firms of implement makers in the county? In general these were men who had begun working as blacksmiths, wheelwrights, millwrights and carpenters and then applied their skills in fabricating metal and wood to produce what the farmer needed for particular tasks in the field and barn. Iron and brass founders, usually located in the towns, also commonly entered the business of agricultural machine and implement making. Some of these town-based foundries had their origins in the eighteenth century and are the earliest Lincolnshire implement and machine makers on record.

Among the hundreds of names listed as agricultural implement or machine makers in the nineteenth century trade directories, a tiny number of Lincolnshire firms - those fortunate enough to be headed by men of outstanding ability and ambition, supported perhaps by significant financial backing - grew from small beginnings to become huge businesses, providing employment for hundreds or even thousands of men, and shining brightly on the international stage. By comparison, modest numbers of firms became implement and machine makers on a noteworthy scale, some gaining a regional or even national reputation, and they developed businesses that survived for a considerable time and provided employment for at least a score or more of workers. At the bottom of the scale hundreds of men were described as agricultural implement makers for only short periods of time before disappearing entirely from the pages of the directories, or perhaps being mentioned later trading as blacksmiths, wheelwrights or iron founders. This is the normal pyramid of success in any sphere of human activity: huge numbers register small achievements; a modest proportion is quite successful; a mere handful reaches the top of the tree. Such is the picture of the implement and machine makers and iron founders of Lincolnshire.

This book does not set out to record the stories of the major Lincolnshire agricultural engineering firms: Clayton & Shuttleworth; William Foster; Richard Hornsby; William Marshall, Robey & Co; Ruston, Proctor & Co. Each of these relative giants deserves a volume to itself, and, over the years, publications in various styles have dealt with the histories of some of these companies *(Book List p.153)*.

The firms featuring in this book are generally from the middle ranks of the pyramid of achievement and all enjoyed a reasonable degree of success. Each began on a similar small scale making simple implements for a local market. But they grew at very different rates and, for reasons of circumstance and opportunity, headed in quite different directions, making a bewildering variety of products, not all of which were purely for the agricultural market.

Perhaps surprisingly, only two of the firms featured here were based in Lincoln (although several more might have been included); the spread of locations represented, from Stamford and Boston in the south to Winterton and Brigg in the north, emphasises the truly county-wide nature of the agricultural engineering industry in the county.

From the relatively small but tenacious T & J Fletcher of Winterton, still making implements successfully today after more than 150 years, to the large and diverse firm, Blackstone & Co of Stamford; from the very adaptable John H Rundle of New Bolingbroke, now noted for making and repairing fairground rides; to Charles Louis Hett of Brigg, who made his name through the manufacture of pumps, turbines and other machinery for water supply and drainage; through two pioneering steam engine manufacturers; this volume has a wealth of information - and hitherto unpublished illustrations - about some enterprising Lincolnshire engineers and their families.

The range of products and services offered is eye-opening and gives a real insight into the manufacturing life of Lincolnshire during the nineteenth and early twentieth centuries. The farming 'revolution' had an impact in all corners of this county, in the towns and villages as well as in the countryside, but the story of the men and women who provided the farmer's tools and machines has not been told in any detail. It is hoped that this volume will go some way to filling this gap.

CHAPTER 1

BLACKSTONE & CO LTD OF STAMFORD

ENGINEERS AND MACHINE MAKERS

Michael Key

Henry Smith

The origin of Blackstone & Co Ltd can be found in the small workshop set up in Stamford's Sheepmarket in 1837 by Henry Smith (1815-1859). Henry and his brothers Nathaniel, William, Robert and Charles were sons of William Smith of Kettering, Northamptonshire, the proprietor of the Lion Iron Works.

Henry Smith first came to public attention with an advertisement in the *Lincolnshire, Rutland and Stamford Mercury* on 29 July 1842, for a newly patented horse rake.

Fig 1
Henry Smith (1815-1859)
founder of Smith and Co, forerunner of Blackstone & Co Ltd.

in the local newspaper that Grant sued Smith over the alleged infringement. Judgement was given against Grant, with Smith receiving costs.

The success of Smith's horse rake encouraged an expansion in the business, which by now included brothers Robert and Nathaniel. Soon a wheeled hand rake and a chaff cutter were patented. Drills for corn and manure (fertiliser), a horse hoe and a moulding plough were added to the firm's catalogue. The range had grown enough for Smith to need both more men and larger premises.

PATENT HORSE RAKE
Registered by Royal Permission

HENRY SMITH'S
Newly-invented Balance HORSE RAKE,
made and sold at his Machine Manufactory,
Sheepmarket, Stamford.

H. Smith respectfully calls the attention of the agriculturalist to his newly-invented Balance Horse Rake, which for its simplicity of make and movement, far excels any that have been made. In this Rake there are neither complicated levers nor chains to raise the tines, they being elevated on a counter balance.

Smith and Ashby

In 1844 Henry Smith joined in partnership with Thomas Woodhouse Ashby (1806-1870). With his new partner Smith acquired the former builder's yard of Thomas Pilkington at the west end of St Peter's Street. This area of land had lain derelict since a disastrous fire in March 1838, when the yard and several buildings on its eastern edge had been destroyed.

Fig 2
Smith & Ashby's patent horse rake, 1859.
The makers boasted that 28 prizes had been awarded for the rake, including one at the Great Exhibition of 1851.

On the success of this machine was based the future prosperity of Smith's young business. However, another Stamford engineer and ironmonger, Joseph Cooke Grant, who had a factory on the bank of the river Welland, felt that Smith's machine infringed the design of his own patented horse rake. Although no records now survive, it is clear from an advertisement

Fig 3
Smith & Ashby's eighteen-inch safety chaff cutter, 1852, the largest of four models they made. HRH Prince Albert bought two for the Royal stables at Windsor.

In October 1845 Smith and Co, as the firm was now styled, opened the 'Rutland Terrace Iron Works, Stamford: New Iron and Brass Foundry and Agricultural Machinery Manufactory'. This rather grand title reflects that it is, or was, next door to an elegant terrace of eighteenth-century houses in Tinwell Road. One can imagine the outcry if such an enterprise was proposed in modern Stamford!

In 1851 Smith and Co sent three of their patented machines, a haymaker, a horse rake and a chaff cutter, to the Great Exhibition of Works of All Nations in London. Each was awarded a bronze Prize Medal. Perhaps not as prestigious as the Great Medal won by Hornsby and Son of Grantham, but it put Smith and Co in the same company as Clayton and Shuttleworth of Lincoln and Tuxford and Sons of Boston, eminent engineering firms all.

The medal was the first of many that the firm and its successors were to win at shows and exhibitions throughout Britain and Europe. To be able to advertise the fact of having exhibited at the Crystal Palace and to have won a prize at the Exhibition was almost to guarantee a growth in sales. In May 1851 the local newspaper reported that Smith and Co 'have received many foreign orders'. Again, in August, 'an abundance of foreign orders', including one for sugar cutters to 'the Netherlands Sugar Refinery of Amsterdam', was noted.

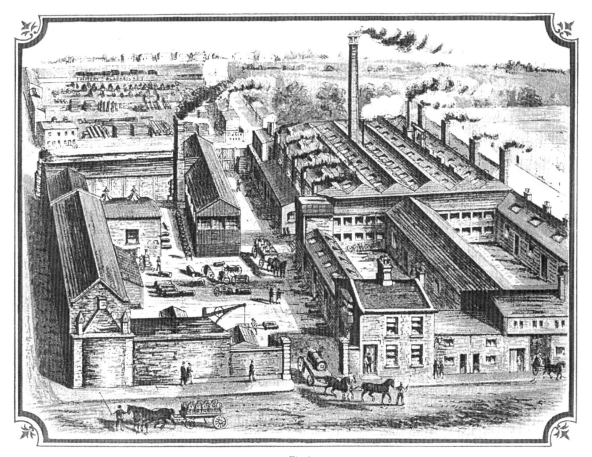

Fig 4
Jeffery & Blackstone's Rutland Iron Works on St Peter's Street, as illustrated in their 1884 catalogue. At the front are the company offices.

In May 1851 Smith and Co took the title Smith and Ashby. In January 1853 Nathaniel Smith moved from Stamford to Thrapston in Northamptonshire, where he became manager of W L Fisher's Nene Side Iron Works. By this time the company's product range had become quite extensive. Apart from the chaff cutters and haymakers, which almost from their introduction had been regular prize winners at the various agricultural shows around the country (a point the firm never failed to mention), there were several types of mills for grinding bones, grain and linseed, and also a large range of cast-iron items including gates, fencing, hurdles, plates, cisterns, mangers, racks and troughs.

For the first time 'the manufacture and repairing of Stationary and Portable steam engines' were mentioned. And the horse rake, which ensured the success of the business, had now, according to the advertisement, sold 972 units and was still priced at £6.10s.

The St Peter's Street premises had gradually been expanded since they were first opened in 1845. Ten years later, in March 1855, new offices and a large showroom, said to be one hundred feet long, were opened *(Fig 4)*. To mark the occasion the firm put on a grand musical soiree for the employees, their families and friends - in total 170 people. Afterwards the firm's newly formed band - forerunner of the popular Foundry Band - entertained the party.

In 1858 Smith and Ashby leased Bonney's Paddock, an area of land between the works and North Street, from the Stamford Borough Council for £12 per year. The 14-year lease had among the clauses controlling the firm's activities one requiring that they install 'a smoke consuming apparatus to any steam engine or furnace'.

Henry Smith died suddenly on 14 October 1859, leaving his widow, Jane, with two very young children. He was only 44 years old. Henry's will stated that Jane and his brother Robert could either continue the partnership with Thomas Ashby or terminate it. The latter course was taken and it was agreed that Ashby should purchase all Henry's models and patterns.

With the Smith/Ashby partnership over, Robert Smith left the firm. In January 1860 he moved from Stamford to Thrapston, where he joined his brother Nathaniel to form Smith Brothers. This company later became the Smith & Grace Screw Boss Pulley Company, surviving until 1995. As a memorial to Smith, Stamford Borough Council erected a public drinking fountain, funded by his executors, next to All Saints Churchyard.

Meanwhile, Thomas Ashby, now trading as Ashby and Co, late Smith and Ashby, was continuing to expand the business. According to the 1861 census the firm now employed 156 hands. Prizes for the quality of their work were still being won: gold medals for the chaff cutter and haymaker at the Great Paris Show, two silvers for chaff cutters and a bronze for a horse works at Antwerp, silver for the haymaker at Sligo.

For the first time the company produced a printed catalogue of the range of products. One, dated 1 September 1860 and inscribed 'No. 2', has survived, entitled 'Agricultural Machinery and Implements manufactured and Patented by T W Ashby & Co, late Smith & Ashby'. In it were described the haymakers, chaff cutters and rakes, also threshing machines and mills.

Also included in the catalogue were the firm's portable steam engines. They ranged in power from 2 hp: price £70; 3 hp: price £80; 4 hp: price £110, and 4½ hp: price £120. All were practically identical to each other. The 2 hp version could be had without wheels as a stationary engine. They were small engines, capable of passing through 'an ordinary four-foot doorway'. Judging by the illustration, the boiler height was no more than a man's shoulder, say five feet. The only useful measurement given is the 4-inch diameter of the cylinder on the 2 hp engine *(Figs 5 and 6)*.

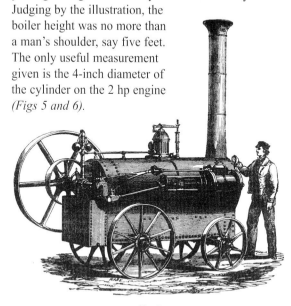

Fig 5
Four horse-power portable steam engine by T W Ashby & Co, 1859. It could burn wood, coal or turf and had a 22-inch pulley fitted to the main shaft to drive barn machinery. Price £110.

No records survive of how many of these engines were made. One similar, apparently converted to a traction engine, was exhibited at the 1871 Smithfield show. The engineering press was unimpressed. *Engineering*

Fig 6
Smith & Ashby portable steam engine of 1850s photographed c1890 at Smithy Fen, Cottenham, Cambridgeshire.

ridiculed the engine saying that it was 'remarkable for little else than the boldness of design bordering on the foolhardy - the pinion being at least 14in distant from any bearing whatsoever.' Not much of an endorsement.

Thomas Ashby had been in the business since 1844 and was described in the 1861 census as a leather merchant, iron and brass founder and agricultural engineer. However, he was probably mainly concerned with the financial side of the firm. Henry Smith's death in 1859 and the leaving of Robert Smith shortly after had left him without an experienced practical engineer. To remedy this he took into partnership in October 1864 George Jeffery 'whose extensive experience and practical knowledge eminently qualify him for the position - in general superintendence of the manufacturing department'. George Edward Jeffery (1837-1889) was born at Yalding, near Maidstone, in Kent. The circumstances of his becoming a partner are that he was probably already working for Ashby. He married Ashby's daughter Mary in 1865.

In 1866 Robert Luke (1839-1909), the firm's bookkeeper, also became a partner. He was a son of Stamford's Inland Revenue Supervisor. Ashby died in 1870 and in 1876 Jeffery and Luke dissolved their remaining partnership. Luke moved to Manchester to found Luke and Spencer Ltd, emery wheel manufacturers.

So, in just over ten years, the firm's title changed three times: Ashby, Jeffery and Co (1864-1866), Ashby Jeffery and Luke and Co (1866-1876), G E Jeffery and Co from January 1876 until June 1877. Jeffery then entered into partnership with Edward Christopher Blackstone (1850-1916), a young engineering draughtsman from London. Now the firm's title changed to Jeffery and Blackstone and Co.

Jeffery and Blackstone

Edward Blackstone was the son of Joseph Blackstone, MRCS (1796-1874), who had moved from Beverley in Yorkshire to set up his practice in London's Camden Town. Edward was educated at Kings College London and served an apprenticeship with the hydraulic engineers J and H Gwynne of Hammersmith.
What persuaded Blackstone to take an interest in an obscure engineering works on the edge of the Fens is unknown.

George Jeffery retired in 1882 to live in Ryhall, a small village about two miles north of Stamford. The reason for his retirement at the early age of forty-five is not known. It may possibly have been a result of a serious fire at the St Peter's Street works in January that year. Despite the retirement of George Jeffery, the firm retained its title of Jeffery and Blackstone and Co for another six years.

By 1886 the St Peter's Street works were becoming increasingly inadequate for the business. Also the company's lease on the site was due to expire the following year. It was therefore decided that a site for a completely new factory should be found. Ten acres of land owned by the Marquis of Exeter adjacent to the Great Eastern Railway's Stamford to Essendine branch line on the eastern edge of the town was chosen. Six acres were purchased and a further four acres were leased from the Marquis, a very significant deal, as this was the first land outside the Borough on which the Marquis had allowed any form of development.

Building work began at the end of August 1886, with the first factory buildings being completed by the following spring. Blackstone himself undertook the design of the factory and the layout of the machinery, a task he was well qualified to do, having been for a time resident engineer with Charles Nelson and Co, cement manufacturers near Rugby, where he was involved in the design and erection of buildings and machinery. The transfer of plant and machinery from St Peter's Street to the new site on Ryhall Road took about a year.

The range and variety of machinery produced is well illustrated by the new catalogue issued by the company in 1887. Listed were corn mills, saw benches, haymakers, horse rakes, harrows, land rollers, chaff cutters, cake breakers, vegetable pulpers and horse gears. Pride of place was held by the portable steam engines.

There were single-cylinder portables from 4 nhp to 12 nhp, the basic engine costing from £130 to £230. The double-cylinder type from 10 nhp to 20 nhp cost up to £245. Also shown was a range of vertical engines. Those with a single-cylinder from 1 nhp up to 12 nhp cost from £45 to £210. The double-cylinder verticals ranged from 10 nhp (£210) to 14 nhp (£270). Whilst these engines differed little from those made by other manufacturers, they show that lessons had been learned after the scorn shown about the company's earlier designs.

In addition to the standard range of engines there was one of unique design, the 'Viator'. This was a portable vertical single-cylinder engine with the boiler mounted between a pair of tall wheels. It was light enough to be hauled by a single horse. The 'Viator' made its debut at the Royal Agricultural Society engine trials held on Gosforth Moor near Newcastle in 1887 *(Fig 7)*.

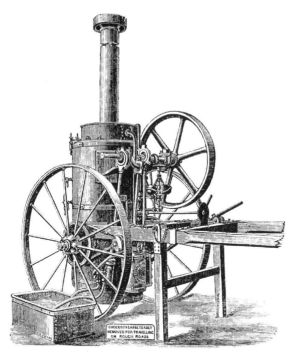

Fig 7
The 'Viator' portable vertical single-cylinder steam engine of 1883. It was available in four sizes from 2 hp to 6 hp and was claimed to be 25 to 30% cheaper than a corresponding horizontal portable engine.

Thirteen engines from seven manufacturers took part in the first such trials organised by the Society since 1872. Jeffery and Blackstone's entry was commended by the judges: 'The finish of the whole machine is good, and during the run of the brake it worked very steadily and well. By the addition of a second cut-off slide and lagging the boiler, the economy might be greatly augmented. Of its class it is a praiseworthy little machine.' Fortunately one of the 'Viator' engines, a 4 nhp type E, has survived in Australia where it occasionally works 'well and steadily.' Sadly, no production figures of the company's steam engines have survived, so it not known how many were made.

A photograph in the Stamford Museum's Blackstone Collection shows a vertical steam engine coupled to a corn mill, elevator and dresser. This is one of a pair, the other showing the same machinery coupled to an oil engine *(Fig 8)*. Both specially-posed photographs are dated 7 June 1900. Whether this was the last steam engine built by the firm is not known, but the photograph suggests that Blackstones still had the capability to build one at that date, should they receive an order.

In order to raise capital following the move to Ryhall Road, a limited liability company was formed on 29 January 1889. The old firm was renamed once more, this time Blackstone and Company Limited, a name it held until the 1930s.

Despite having built the new Ryhall Road works, Blackstone seems to have needed more space and in 1892 acquired Newboult's Rutland Plough Works at Great Casterton, a small village about two miles north-west of Stamford. During the same year Blackstones also acquired All Saints Iron Works in Scotgate, Stamford. Exactly where these two works were situated is uncertain. The former was probably on the right hand side of the old Great North Road, just before the road to Little Casterton. A terrace of cottages now stands on the site. The latter probably stood at the northern end of St John's Terrace, a small lane which runs parallel to, and behind, the houses on the right of Casterton Road as one travels north. There appears to be no record of how long the firm held these sites.

The First Oil Engines

During the middle years of the nineteenth century a powerful new rival to the steam engine came on the industrial scene when the internal combustion engine made its appearance. This is not the place to describe the development of these engines; suffice to say that by the early 1880s gas and oil engines were introduced as reliable power sources and numerous applications were found for them. The new engines were generally simple to operate and, provided fuel, lubricating oil and cooling water were available, their maintenance was within the capabilities of most people.

Fig 8
A 'Colonial' flour set: mill, elevator and dresser driven by a continuous lamp engine in 1900.

Like many other engineering concerns, Blackstones were attracted to this new and potentially profitable product, and began to look for a suitable design to manufacture. Tom Price, then the firm's head sales representative, later claimed he was told by Edward Blackstone to 'find a good engine and somebody who knows all about them.' Price, who regularly attended the many agricultural shows around the country, found the combination required by Blackstone at the Darlington Royal Show in June 1895 in the form of the Carter Brothers of Billingshurst, West Sussex, and their 'Reliance' oil engine *(Fig 9)*.

When the Carters built their first oil engine is unknown, but by May 1894 they had made their first patent application: 'No. 9889: Improvements in and Connected with Petroleum Oil Engines'. Like several other makers, the Carters used a tube vaporiser on their engine. This was difficult to keep hot enough to vaporise the oil and air mixture unless a continuous flame was applied. Their patent describes four methods of directing the engine's exhaust gases through a passage surrounding the vaporising tube. It still had to be started with a blowlamp to heat the ignition tube, but, once the engine was running, the action of the exhaust kept the vaporiser at a sufficient temperature. In practice this was less efficient than had been hoped. However, in 1894 they put their engine on the market as the 'Reliance'. A second patent was obtained in 1896 for a much-needed improved vaporiser. The inlet valve now had proper valve guides and the oil inlet pipe had a one-way valve for the regulation of fuel.

In a later licensing agreement, dated 18 September 1896, E C Blackstone acquired the sole right to exploit the Carters' patents and purchased all the Carter Brothers' oil engine patterns, castings, parts and fittings for £110. One may think that was a sound investment. Under the agreement Frank and Evershed (Tod) Carter were to move to Stamford to develop the Carter engine. As licensee, Blackstone was to pay a royalty of ten shillings per horse-power on each engine in which the Carter patents were used as they left the works, whether they were sold or not. Payment was to be not less than fifteen shillings for any one engine. These terms would be renegotiated in due course. All royalties were to be divided equally between the four brothers, with a minimum of £50 to be paid in 1897.

Fig 9
The prototype Carter Brothers 'Reliance' engine, forerunner of Blackstones' successful oil and diesel engines.

Fig 10
A Blackstone swath turner manufactured in 1904, with horse shafts removed, a rare survivor of the thousands produced by the firm.

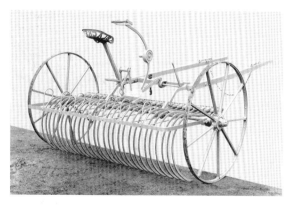

Fig 11
This 1900 Blackstone horse-drawn hay rake has changed little from the original one patented by Henry Smith in 1842.

Fig 12
Blackstone haymaker of 1890. According to the company's 1937 catalogue of haymaking implements, over 4000 of these machines had been sold.

In subsequent years this was raised to £70. In 1910 the Carters gave up all claims to any share in the profits of the company and any claims to royalty monies. Since August 1897 the four Carter brothers had each received £3,500 after tax.

By the exploitation of the Carter patents, Blackstones were to become one of the country's leading engine builders. The first engines left the Stamford works in October 1896. The first, a 4 hp carrying number 10497, was sold to McCowan of Tralee. The second, number 11543, 1 hp, went to Keightly of Birmingham. No more left the works until February the following year. If consideration is given to the time it would have taken to set up the works to produce the new engine, it is probable that these two engines were in fact Carter Reliance engines.

For what, even then, was a very important step for the company little appears in their records. In March 1897 it was resolved to increase the capital to £34,000 as 'the company has recently taken up the manufacture of oil and gas engines.' In 1898 E C Blackstone reported 'that in consequence of the increasing demand for oil and gas engines it was necessary to enlarge the shops and put down new machines to facilitate the manufacture of these articles.' There followed a rapid decline in the manufacture of steam engines. As mentioned above, the last was probably built in 1900.

There was no such decline in the manufacture of barn machinery and agricultural implements, nor of innovation in these products. The first swath turner had been built in 1892 and an improved model was introduced in 1900 *(Fig 10)*; a new land roller and pulveriser in 1904 and a new self-acting horse rake in 1905 *(Fig 11)*.

But it was on the engine side that the most innovative progress was made. A new vaporiser and igniter were introduced in 1903; the combined engine and water pump in 1906; an oil-engine traction engine in 1907; and a sectional engine in 1908. The most important step forward, however, was the introduction of Blackstones' first hot-bulb dual-ignition oil engine in 1908. In this, a wide range of heavy oils and residues could be used as fuel. A series of small vertical engines was introduced in 1913. In three sizes, 1 hp, 3 hp and 5 hp, these four-stroke engines were to remain in production until about the end of 1915.

Sometime before 1913 Blackstones bought a group of old cottages and a shop in Stamford's Broad Street, and there established what they called the 'Town Offices' and retail department. During 1913 these buildings were demolished and a modern showroom with offices above was built on the site.

Erected by a local builder, William Woolston, it was a spacious building with large display windows. This elegant building was sold in 1925 to the Cambridge Cinema Company which, during the following summer, transformed the former showroom into the luxurious Central Cinema. The seven-hundred seat auditorium was reported to have cost £14,500. Sadly this building was gutted by fire in May 1937 and had to be demolished.

The First World War and the Inter-war Years

Shortly after the outbreak of World War I the Rutland Iron Works became a government-controlled establishment. Production was turned over to the manufacture of munitions, submarine engines and spare parts for naval motor launches for the Admiralty. A special plant was designed and installed for the manufacture of 6-inch and 18-pounder shells. Women were recruited as machinists to replace men called up for the Services and the skilled workers who were sent to work in the shipyards of Newcastle on Tyne *(Fig 13)*.

Edward C Blackstone died in 1916 after a short illness. For much of his time in Stamford he had lived in Rock House, an elegant early Victorian building surrounded by attractive gardens and shrubberies in Scotgate, but he moved to a newly built house in Peterborough a few years before his death. He was said to have been a good and caring employer, encouraging many social and sporting activities for the men and women. An annual works' swimming gala was held in the nearby river Welland. The Blackstone football team, the 'Stones', was founded during his ownership.

In 1919 Blackstones became part of AGE, a group of agricultural, transport and engineering companies, mainly located in eastern England. Although this included substantial and hitherto successful firms such as Aveling & Porter Ltd and Barford & Perkins, poor financial management and the very difficult trading circumstances of the inter-war years brought the combine into liquidation by the end of 1933 with debts exceeding £2.5m.

This venture was a financial disaster for Blackstones, as indeed it was for the other participants; nevertheless there were some interesting and important developments and co-operations during that period. Among the more profitable was the introduction of a centrifugal pump. In a licensing agreement drawn up

Fig 13
Women munition workers at Blackstones during the First World War. In the centre at the rear stands the bald-headed George Blackstone, a son of E C Blackstone, the founder.

in 1923 with the Unchokeable Pump Company Limited, Blackstones gained the right to produce that company's pumps for the next twenty years. By a further agreement, six years later, Blackstones acquired full ownership of the pump company's patents.

The 'Unchokeable' pumps were capable of moving hard or soft solids in fluids and semi-solid materials. The liquid or semi-solid mixture was forced through the pump casing between two revolving plates that formed part of the impeller, not part of the casing side plate. Two diametrically opposed impeller ports allowed the unrestricted flow of material and gave rise to the pump's uncompromising name. They were usually driven by a direct-coupled electric motor, but models for belt or direct drive from diesel or petrol engines were available.

The ability of this pump to deal with a wide variety of materials soon found a ready market with municipal authorities, who installed them to handle treated and untreated sewage. Many more were set to work in sand and gravel pits, whilst others were to be found pumping slurry in quarries or aboard sea-going dredgers. Building contractors used small 'Unchokeable' pumps for dewatering trenches. Just before the Second World War, Blackstones began a withdrawal from the pump market, but production of the pumps did not come to an end until early in the 1960s.

From soon after the First World War the Carter brothers began working to develop an improved form of fuel injection system for their engines. Their work to replace the so-called hot-bulb ignition that they introduced in 1903 culminated with the introduction in 1924 of the 'Spring Injection' mechanism.

Simple in both design and operation, it allowed a measured amount of fuel to be pumped to the injection chamber at a precise moment and for sufficient time for the fuel to be fully burnt. With close control of fuel consumed and an ability to start from cold and to run on practically any fuel, the company claimed the spring injection engine could successfully challenge steam in its last stronghold, the power-station steam turbine.

Fig 14
*Blackstone agricultural implement assembly shop. In the foreground is a batch of haymakers;
behind them side-delivery rakes and swath turning rakes.*

Fig 15
The light machine shop in the early 1920s. The uneven timber floor was swept away when Lister and Co took over in 1936.

The first engine with the new fuel system provided power and lighting for the Palace of Engineering at the British Empire Exhibition held at Wembley in 1924 and 1925. The new system also enabled Blackstones to produce their first high revving diesel engine, the BHV designed by Frank Carter. This pre-dated by six years the Gardner Brothers famous LW engine, which was designed specifically for automotive use, and, as history shows, with more success. The BHV, which developed 10 bhp per cylinder at 1000 rpm, was used experimentally in several road vehicles, usually elderly lorries, a 1916 Thornycroft and an early Pierce-Arrow among them. Reliable performance figures are hard to find, but there does seem to have been considerable economy over similar petrol engines.

To meet the demands of the small power user, Blackstones introduced the so-called 'No Trouble' petrol engine in 1922. This compact, reliable and simple vertical engine was ideal for use in workshops, driving air compressors, small electrical generators and the like. It was originally designed as a side-valve engine developing 5 bhp, but grew into a range of four sizes of overhead valve engines from 2 to 7 bhp, running on either petrol or kerosene. These stayed in production until 1935. The title 'No Trouble' only applied to the 5 bhp side-valve engines and was not given to the overhead valve engine.

Under the Carters' influence, Blackstone & Co were often involved in unusual and innovative designs. A curious development of the vertical petrol/kerosene engine was the 'Fuelol'. In 1931 the Carters took a 5 bhp engine and adapted it to run on fuel oil, hence the rather awkward name. It was only moderately successful, but it paved the way for the later highly successful DB.

At the other end of the power scale was the mighty TN8. This eight-cylinder vee engine was built in 1931 for the steel rolling mills of Pemberton Tinplate Co Ltd of Llanelli, Wales. Its job was to drive a four-mill rolling plant with a 110-ton flywheel at 36 rpm continuously for five and a half days a week (126 hours), twelve months a year. In addition, it had to provide all the electric power for driving hot and cold rolls, shears and bar cutters, the cranes and the lighting. For this the Carters produced what was to be their largest engine. With 18-feet long, four-throw crankshaft that could withstand the overload shocks, cylinders of 18-inch bore and 24-inch stroke, the engine developed 1,250 bhp at 250 rpm. Carters' spring injection system enabled the engine to maintain output regardless of load and to work efficiently at low speed. This pioneering engine was a success and a second, four-cylinder version was built for Swansea Steel Products in 1935. It is thought these engines were scrapped in the 1950s when the Welsh steel industry was going through a major shake-up after nationalisation.

Meanwhile, following the ill-fated AGE combine, the company reorganised and Barclay's Bank became the new owners. Three Blackstone brothers were relegated to the role of minor shareholders, though they each held important management roles.

Frank and Tod Carter's last design was the EPV, which was begun in July 1933. This was a totally enclosed vertical engine with an 8-inch bore and 11-inch stroke developing 40 bhp per cylinder at 600 rpm. It had opposed side valves and swirl combustion chambers, similar to the later horizontal spring injection engines. The enclosed design was neat and elegant; accessibility through the side covers was easy and quick. The EPV went into production in 1934, and the first to be sold, a six-cylinder engine, No. EPV6 189182, was despatched on 16 June that year *(Fig 16)*.

In 1935 under another reconstituted board, Blackstones entered into a ten-year production agreement with Massey-Harris Co Ltd of Toronto by which all the agricultural products of the Blackstone Company were to be sold through a centralised organisation. Massey-Harris was possibly the largest manufacturer of agricultural machinery outside Europe and Blackstones' celebrated haymaking implements had been produced for nearly a century. Both companies were able to draw on each other's expertise in manufacturing and selling. It was agreed that Blackstones' entire agricultural output was to be manufactured to the specific requirements of, and sold through, the Massey-Harris organisation. From then until 1944 all agricultural machinery produced at Stamford carried the title 'Massey-Harris-Blackstone'. There is no doubt that this arrangement allowed Blackstones to stay afloat during a very difficult period.

Lister Blackstone

1936, however, saw the end of Blackstones' independence when R A Lister & Co Ltd of Dursley took a majority shareholding in the company and installed C Percy Lister as Chairman. Founded in 1867, Listers produced an eclectic range of agricultural machinery, dairy equipment, sheep shearing machinery and small petrol and diesel engines. However, Listers wanted to get into the potentially profitable large engine market, and the acquisition of Blackstone & Co in 1936 gave them that entry. Lister and Blackstones' decision was a sound one. The next thirty years would possibly be the Stamford firm's most prosperous period. (Curiously the company's name during this period can be found as either Blackstone & Co Ltd or as Lister Blackstone & Co Ltd. There appears to be no set rule for this, although agricultural implements were usually inscribed 'Lister Blackstone', at least after the Massey-Harris agreement ran out. For the purpose of this narrative the title 'Blackstone' will continue to be used).

The last connection with Carters was also severed at this time. 'Tod' Carter, the firm's Chief Engineer, retired when Listers took over. His brother Frank, the Works Manager since 1904, had died in 1934. Between them they had had a profound effect on the development of the internal combustion engine. Among their achievements were the vaporizing engine

Fig 16
The controls end of an eight-cylinder EPVC marine engine. This was the last engine to be designed by Frank Carter, shortly before his death in 1934.

Fig 17
Rutland Iron Works, Ryhall Road, Stamford in the 1930s. The white-fronted rectangular building with the flat roof is the tractor shop, which, when erected in the late 1920s, was the first pre-stressed concrete building in the East Midlands.

with automatic ignition, the camshaft governor, the spring injection system and the 'cCc' (Carters' compound cam) injection system used on the high-speed general-purpose engines.

The period immediately following the take-over was one of intense activity. One of the first of Listers' tasks was improving and modernising the Stamford factory. The internal tramways were uplifted and the roadways resurfaced for a number of Auto-trucks. In the machine and erecting shops the old sand and tarmac floors were concreted. The overhead line shafts which drove the machinery were removed and an electricity power supply installed for each machine. With these changes the workmen had a much cleaner and lighter environment.

Tod Carter's successor as Chief Engineer was Percy Jackson, who joined the company from Petters of Yeovil. He took up the development of the EPV, producing a range of industrial and marine applications in 2 to 8-cylinder variations. After Listers took over, the Carters' fuel injection system was dropped in favour of Bosch equipment, giving them access to the world-wide service facilities offered by the German company.

Listers had originally acquired Blackstones in order to gain access to the large engine market through the EPV engine. However, they also needed to expand and modernise the range of small engines. The result was the P type range of totally enclosed horizontal engines.

The first P type to go into production, in 1937, was the OP. This had a bore and stroke of 7 inches by 9 inches. Running at 500 to 700 rpm, it developed 18 to 26 bhp. The first production OP, No. 196515, was despatched in October. In February the following year the JP was introduced. Smaller than the OP, with a bore and stroke of 5 inches by 7 inches, the JP developed 10 to 16 hp at 500 to 800 rpm. This range of engines was expanded to the RP and SP (introduced June 1938), the TP (May 1939), and the MP (July 1955). There was also a twin-cylinder RP. The popularity of the P type engine was such that a line erection track was installed 1949 to keep up with demand. Component stores were arranged alongside the erection track to keep up a continuous supply of parts.

By the end of 1954 production of the P type averaged nearly 600 engines per year. Most of these engines were sold abroad for all kinds of applications. The majority went to the Middle East driving irrigation systems. Production of the P type continued right up to the end of engine production at Stamford in 1993, but even then large numbers were supplied as knocked down kits.

In February 1939 Blackstones entered into a five-year sales agreement with the Brush Electrical Company of Loughborough. This was to sell the M range of horizontally opposed engines, to be designated 'Blackstone-Brush'. These were very low, compact multi-cylinder, low-revving engines suitable for industrial and marine applications. In size they ranged from the four-cylinder 4M9 rated at 200 bhp to the sixteen-cylinder 16M13 rated at 1,520 bhp. One of their selling points was their low height - seven or eight feet - which made them particularly suitable for marine installations in between decks.

The Second World War

The coming of the Second World War gave the impetus for a radical rethink on the types of engines and machinery produced at Stamford. All the open-crank spring-injection engines, the small vertical petrol, kerosene and diesel engines and the BPV automotive engines were dropped. Only the DB, P and E ranges of engines were kept in production, all three in large numbers. In particular, the production of EPV engines for base-load installations was increased to meet the demand from all three Services. Stand-by generator sets powered by EPVs were installed in purpose-built trailers which were to be located at strategic points for use if the main power supply was destroyed by enemy action. They were also used to supply power for RAF Barrage Balloon units and Royal Observer Corps searchlight units and listening posts. At Plymouth the Royal Navy radio communication network and Radar station were powered by an EPV. A four-cylinder version of the engine was produced with an integral gearbox to go into small wooden-hulled minesweepers. A large quantity of DB engines, painted grey, was supplied to the Air Ministry. Small bore 'Unchokeable' pumps, direct-coupled to Lister petrol engines, were mounted on trailers for the Civil Defence and National Fire services.

The agricultural department was not slack either. The demands for increased productivity from the land produced a similar demand for agricultural implements of all kinds. Large numbers of hay rakes and haymakers were built and it was claimed that around 5,000 potato lifters were made. The production of stone and plate grinding mills and kibblers, however, had ended before the war began. There was, apparently, still one elderly craftsman at this time who could dress the millstones, but this service came to an end on his death.

Because of the shortage of wood, the company was unable to meet the demand for hay and straw elevators. So in 1941, having gained permission from the Ministry of Supply, a steel version was designed. Any wood that was available was being used to make ladders for the fire services.

Apart from the engines and implements, the Stamford factory turned out an extensive range of unrelated products. Among them were instrument stands for searchlight units and other uses, and breech blocks for two-pounder anti-tank guns. Slewing rings for tank gun turrets were turned on an old lathe that had formerly been used to turn ten-foot flywheels. Gearboxes for shell manufacturing lathes were built for Alfred Herbert Ltd of Coventry, who was then the country's leading machine tool manufacturer.

Before the war only a few women worked for the firm, either in the drawing office as tracers or as general office clerks, but now that men were being conscripted more women were brought onto the factory floor. The company's war effort was not confined to engines and machinery. There was also the important morale-boosting exploits of the Blackstone Social Club, the locally produced newspaper *News of the Lads*, and the Blackstone Follies, a talented group of six men and women who put on concerts in the works canteen and at local army and air force units.

Although the Blackstone works was known by the German Luftwaffe to exist (an aerial reconnaissance photograph in Stamford Museum confirms this), there was no serious attack on the works throughout the war. Two vague attempts were made, but both did little damage - broken windows mainly - and no serious injuries were caused.

Fig 18
Corn mill with French burr stones, c1885.
This was one of the smallest models in the range.

While the factory was concentrating on its war effort, research and development of new and improved designs continued. Percy Jackson had taken over as Chief Engineer in 1937 and George Hallewell as Chief Draughtsman in 1935.

By 1944 Hallewell had produced the prototype EV engine. This was a logical development of Carters' EPV and had the same bore and stroke: 8 inches by 11 inches. The EV had a single-piece cylinder housing with an overhead-valve cross-flow head, a crown piston and a stiffer crankshaft and camshaft. The output was initially not much more than the EPV at 45 bhp per cylinder at 600 rpm, but its great benefit was its economy of production. The first EV engines were despatched in November 1950 and became one of the firm's most successful engines.

Nineteen-Fifties and Sixties

Turbo-charging was introduced in 1951, increasing output to 60 bhp per cylinder at 600 rpm. Available in 4 to 8-cylinder versions and designated EVS, the first of this series was despatched in February 1951. By 1953 the normally-aspirated EV had been up-rated to 56 bhp per cylinder at 750 rpm and re-designated ER. By 1954 the turbo-charged version ERS was producing 75 bhp per cylinder at 750 rpm. Both versions of the EV were to be developed further by Hallewell until his death in 1962. However, his work laid the foundation for the development of the E type engine. This engine would find its way into world-wide applications: ground-base installations (power generation), marine propulsion and railway locomotives.

With the ending in 1944 of the ten-year agreement with Massey-Harris for the supply of agricultural implements, the name now appearing on the company's products became 'Lister Blackstone'. During this early post-war period many hundreds of implements - hay harvesters and potato lifters in particular - were produced. Most went to the Continent to replace machinery destroyed during the war and to help increase food production across Europe.

Lister Blackstone was acquired by the Hawker Siddeley Group on 1 June 1965 and merged with engine builders Mirrlees National Ltd of Stockport four years later. The company name was then changed to Mirrlees Blackstone Ltd. Hawker Siddeley's roots were in the aircraft industry and it was a mere newcomer to the world of engines. The new firm was able to develop its current range of engines to the limit

Fig 19
Turnip or root pulper, c1900, which has been restored with new leg assembly and fresh paint finish. It is driven by steam or oil engine.

of output, but the new structure and changed corporate priorities prevented the evolution of new designs.

The 1960s was a good period for engine manufacturers. New markets were opening in the developing world and Blackstones were able to take advantage of them. However, increased production, particularly of the larger engines, and the imposition of more stringent test requirements showed up the inadequacies of the existing works at the Stamford site. An extensive renewal programme, which would continue into the 1980s, was put in hand. The foundry, heavy machine shop, erecting and testing shops, and later, the spares and service department, would all be re-equipped. While all this modernisation was taking place other changes were afoot. In 1968 another attempt at a small vertical engine, the BV, was made, but only a prototype was built.

When Hawker Siddeley took over, agricultural machinery production was moved to Listers' works at Cinderford in Gloucestershire. This created a need at Stamford for a second product, and a Gear Division was formed at Easton on the Hill, about four miles west of the town, in Easton House. This is a grand building which in the 1820s had belonged to Henry Shuttleworth of Clayton and Shuttleworth. Blackstones had purchased the house for use as a hostel for overseas visitors and foreign apprentices. Now part of it became Ralph Ley's design office. From here came some very successful gear and gearbox designs. In the 1970s, for example, 2,600 special gear and pinion sets were produced, to Ley's design, for the London Passenger Transport Board when they re-equipped the Piccadilly underground line. Ralph Ley had also produced a design for a new vertical engine, the BV, a multi-cylinder engine with a 7-inch x 7-inch bore and stroke. A prototype was built and successfully tested, but once more the Stockport directors decided not to proceed. Only the existing engine types were to have continuing development.

The Final Phase

During the early 1970s the works was turning out about thirty complete engines per month, many of which went into Scottish fishing boats. Apparently this business was so good that a resident service engineer, fully stocked with spare parts and even spare engines, was permanently based in Aberdeen. His job was to keep the trawler fleet at sea. Then, in 1972/73, the Scottish fishing industry collapsed and with it the engine market in that division. The Middle East had also long been a profitable market, especially for the P type engines, used for irrigation systems. It was said that one could stand anywhere in Iran or Iraq and in the quiet of night hear the beat of a Blackstone engine at work. The 1973/74 Iran/Iraq War closed down this market opportunity.

Despite the earlier decisions not to produce any new engines, in 1976 the compact Vee-configured ESL was introduced in both 12 and 16-cylinder versions. Using the, by then, standard bore and stroke dimensions, 8 inches x 11 inches, this engine was capable of producing 125 bhp per cylinder. 1978 saw the introduction of the E Mk II engine. With the same bore and stroke as the original EPV designed as long ago as 1933 by Frank Carter, it could produce a maximum of 180 bhp per cylinder at 1,000 rpm, nearly five times the output of Carter's engine.

During the late 1970s the Stockport drawing office produced a general design for another completely new engine to be produced at Stamford. Designated MB190, it was a 45° Vee with a 210 mm stroke and 190 mm bore. It turned out to be more complicated and expensive to manufacture than any other Blackstone product. It was built in four configurations: 6 and 8 cylinders in-line and 12 and 16-cylinders Vee.

At 1,500 rpm it developed 173 bhp per cylinder. Although the engine itself was a success, it was too expensive to make and too fast at 1,500 rpm for many of the company's markets. Only about fifty engines were built before it was dropped.

One application which could have saved the MB190 was the use of the 12-cylinder Vee version to re-engine British Rail's HST125, the High Speed Train. These diesel train sets, which ran regularly on the East Coast Main Line, entered service in 1976. After about three years' running, the original Paxman Valenta engines began to suffer a number of problems, including fractured connecting rods and cracked cylinder heads. In 1979 BR approached Mirrlees Blackstone with a proposal for developing the MB190 for rail traction. British Rail insisted that the new engine should directly replace the existing engine with the minimum disruption to the engine layout and equipment. Following trials, using a six-cylinder engine to simulate the installation in the HST, which satisfied both the company and BR of the MB190's potential, BR placed a £750,000 order for four engines. The first complete engine was delivered in April 1986 *(Fig 20)*.

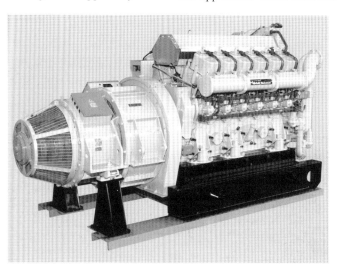

Fig 20
Mirrlees Blackstone MB190 Vee engine designed for use in British Rail's HST 125 trains, but was more successful in power generation installations.

Four locomotives with MB190 engines were put into service on the Western Region of BR, where they performed adequately, but not spectacularly. For a time a Blackstone engineer was based with the engines during the running-in period. It seems that once he left, these engines, being the only ones in the fleet, became neglected and received less attention than the others. No further MB190s were supplied to BR.

The general recession in world trade during the late 1980s and early 1990s, particularly in engineering, hit Blackstones as hard as anyone else. As part of the Hawker Siddeley Group, they were subject to a major reorganisation and lost many of their 650 employees. Because of the loss of many important markets, a fourteen per cent decline in profits was recorded for the first six months of 1991. This resulted in the take-over of Hawker Siddeley by BTR (British Tyre and Rubber Company) in November 1991.

During the winter of 1993/94 engine production was moved from Stamford to Stockport. The last engine left the Stamford works in March 1994. This latest upheaval meant that a large proportion of the workforce was made redundant, although a few were given the option of moving to Stockport. Only the foundry, spares and service departments remained at Stamford, supported by a machine shop with about fifty machine tools. Here the castings were finished and some E range engines refurbished. A market for finished castings for engine manufacturers at home and abroad was developed.

During 1994 the BTR group split up and the Stamford and Stockport sites became a part of the Anglo-French consortium GEC Alsthom. This arrangement lasted about a year until MAN B&W of Duisberg, Germany, took a controlling interest. Then, because of the German economic situation, in 2002 MAN began closing down many of their outside operations. The effect on Blackstones of Stamford was the closure of every department except the Service section, which moved to a small office on a new site.

For one hundred and sixty-five years the company served the town of Stamford as employer and the wider world as manufacturer, for one hundred and six of those years from The Rutland Ironworks in Ryhall Road. The works are no more, but the quality of its products is still valued and celebrated by many people around the world.

Fig 21:
A display of Blackstone machinery dominated by a selection of grist mills on the stand of Messrs Mangold Brothers, Port Elizabeth, South Africa. The Stamford company successfully marketed its engines and agricultural machinery in many parts of the world, especially in the late Victorian and Edwardian periods.

CHAPTER 2

JOHN COOKE OF LINCOLN

PRIZE PLOUGH MAKER

Hugh Cooke

John Cooke's Lindum Plough Works was, between the 1850s and 1920s, Lincoln's largest manufacturer of agricultural carts and wagons and the largest plough makers in Lincolnshire, with business throughout the country and overseas.

The factory was well placed to attract the attention of farmers attending the Cattle Market on Monks Road, being between Rosemary Lane and Montague Street and extending down to Croft Street. A plough was mounted on the Monks Road boundary wall to advertise the business.

John Cooke was born at Eagle in 1821, apprenticed locally and by 1841 was working as a wheelwright in Eagle. In 1854, at the Royal Agricultural Society implement trials in Lincoln, the performance of Mr Cooke's general purpose plough was described as excellent. By 1861 John Cooke was trading from premises at 20 Monks Road, Lincoln as an implement maker specialising in 'Prize Ploughs' and living at 21 Monks Road with his young family.

The Midland Railway connecting Lincoln to Nottingham and Derby had arrived in 1846, followed in 1848 by the Manchester, Sheffield and Lincolnshire and Great Northern Railways, providing access to Yorkshire. Lincolnshire's farmers gained immediate access to the growing populations of South Yorkshire and the East Midlands and, at the same time, Lincoln-manufactured products could be supplied to those growing markets.

The expansion of Lincoln's engineering industries during the 1860s and the development of the Monks Road area started to gather pace, with housing being built down Rosemary Lane and Baggeholme Road, followed by a new road, Montague Street, between Monks Road and Croft Street. John Cooke's Plough Works moved into a new factory, now named the Lindum Plough Works, in about 1870 *(Fig 55, p47)*.

Some engineering companies chose locations adjacent to the new railways to simplify the transport of heavy products to their customers. John Cooke chose the location nearest to his customers at the Cattle Market, his implements being easily carried to the railway station for transport.

In the middle years of the nineteenth-century the farmers of Lincolnshire were actively investing in all means to improve their productivity and were

Fig 22
Hermaphrodite wagons could be quickly converted from a sturdy 2-wheeled cart for carrying heavy goods to a 4-wheeled wagon for cereals or forage. 'Moffreys' were unique to Lincolnshire and the eastern counties.

prepared to adopt new tools and techniques to that end. John Cooke recognised an opportunity in these circumstances. Whilst village wheelwrights and smiths could make competent wagons and farm carts, they lacked the expertise and resources to make high-performance ploughs. John Cooke developed the expertise and invested in the resources to become one of a handful of plough makers competing for business across the country and seeking export business.

An indication of the level of business he eventually developed is given in the *Illustrated Catalogue* of 1876:

> Cooke's 'A' Wood Ploughs are now well known in all parts of the kingdom, but especially in Yorkshire, Lincolnshire, Nottinghamshire, Derbyshire, and all the Midland Counties. Several thousands are now in use, and the average yearly sale in Yorkshire alone is several hundreds. Price £2:10 shillings.
>
> Over 3000 Double-Furrow Ploughs are regularly in use in the Midlands; over 500 were made and sold last year. Price of the Mark XLR: £10.[1]

The Plough Works wage book for 1864 to 1866 shows employee numbers rising from 15 to 24, most working a full six-day week. About one-third of the workforce were skilled men earning over £1 per week, about one-third semi-skilled earning between 10 shillings and £1 and the remainder were labourers earning under 10 shillings. There must have been a considerable demand for skilled workmen at this time of rapid industrial growth in Lincoln and John Cooke was paying wages which appear to be good and competitive.

By the 1870s John Cooke's Ploughworks was making a range of products for the agricultural market, as listed in the 1876 Catalogue:

> 3 models single-furrow wooden framed light-weight ploughs
>
> 5 models single-furrow iron & wood ploughs for medium to heavy duties
>
> 2 models single-furrow iron framed heavy-duty ploughs
>
> 4 models double-furrow iron & wood ploughs for medium to heavy duties
>
> A comprehensive range of plough fittings and wearing parts
>
> Horse hoes for turnips etc.
>
> Lever scarifier or drag
>
> Iron harrows and wood harrows
>
> Cambridge rollers and flat rollers
>
> One-horse carts to carry up to 25-30 cwts; cost £16 10/-
>
> Two-horse agricultural carts to carry 40 cwts; cost £19.
>
> Combined cart and wagon (hermaphrodite); cost £25 *[Fig 22]*.
>
> Railway and merchants' drays on springs and Cooke's celebrated wagons:
>
> Made to order, prices and particulars on application.[2]

Carts and Wagons

Carts and wagons were made by wheelwrights, and that was John Cooke's craft and the trade in which he started in Eagle and then transferred to Monks Road.

It was not until the mid-1840s that any mention is made of his ploughs, so his early business must have been predominantly in carts and wagons. Presumably John Cooke's earliest vehicles were agricultural carts, drays and wagons, followed later, when he was trading in Lincoln, by sprung vehicles for use on urban roads. The catalogues from 1876 onwards provide illustrations of light traps and market carts, contractors' carts, merchants' drays, brewers' drays and delivery carts, as well as the celebrated wagons, farm carts and lurries *(Fig 23)*.

John Cooke's carts tended to be clean and simple in design, without the intricate woodwork of the smaller producers. The carts appear to have been built to order. Stocks of finished carts would have demanded storage space, which is not apparent from the catalogue illustration of the factory premises. Panels and wheels were probably constructed when other work was light; this would speed up the delivery once an order had been received. Ample stocks of the seven types of wood used were held for seasoning in the store area. Farm implements needed to be well finished to be competitive and the finish and presentation of John Cooke's carts was among the best.

The reporter for the *Implement and Machinery Review* of August 1878 complimented Mr Cooke on the superior appearance of his implements and received the reply, 'Well, they ought to look nice, for I personally examine everything before it leaves the Works'.

John Cooke was Lincoln's largest manufacturer of agricultural carts and wagons. They were sold principally in Lincolnshire, where they competed with a large number of smaller, village based producers. Farm cart design differed from county to county, depending on local crops and local preferences. This, together with the cost of transport, probably limited John Cooke's vehicle market.

Fig 23
Cookes illustrated nine types of vehicle for various uses; sprung carts provided greater safety for fragile goods and more comfort for carters hauling goods over long distances.

Prize-Winning Ploughs

The earliest record of Cooke's ploughs being evaluated was in 1854 at the exhibition and trial of implements held by the Royal Agricultural Society in Lincoln. In the report of the Lincoln meeting of the Society it is observed that 'So much has the desire of the cultivators of land increased for improved implements and agricultural machinery that a new class of manufacturer has sprung up - a class second to none in intelligence, perseverance and skill - whose inventive powers are severely taxed to keep pace with the requirements of their customers'.

Thorough comparative trials were organised 'to give the prize to the plough that, running upright and steadily on the furrow-sole, without pitching, tilting, or swerving from such upright position, should cut and turn its work in the best manner, and lay it up in the best form with the lightest draught; the plough itself to be simple in its construction, free from complication, can be kept in order at the least expense, and the original cost fair and moderate'.

The description of the trials of heavy ploughs adapted for ploughing more than 10 inches (25cm) deep shows how tough these tests were. 'The field was of a very heavy and adhesive clay soil. Six ploughs competed in this class, four of which were quite speedily seen to be incapable of standing up to the severe test to which they were subjected; still the anxiety of their exhibitors to prove their capabilities was so great, that ultimately eight very powerful horses were attached to each plough, and the ploughing, if we may so term it, became interestingly absurd; for, in addition to the horses, four or five leaders were to be observed with them, and a similar number in holding and riding upon the ploughs to prevent them being thrown out of work by the tenacity of the soil, and to force them to turn over such an unexampled furrow-slice, frequently comprising a depth of 12 inches by 16'. Messrs Howards' plough got the prize and the Ransomes & Sims plough also survived; the whippletrees and plough harnesses were wrecked. All the ploughs tested at this meeting were single-furrow.

There were fifteen entries in the trial of general purpose ploughs, John Cooke's being one of them. The first trial was on heavy land, using four horses and ploughing not less than 7 inches (18cm) depth. Wooden ploughs were considered to have little chance under such conditions. Nevertheless, Cooke's plough was said to have performed excellently and it joined the seven metal ploughs for the second trial. Here the soil was about 7 inches deep mingled with much stone. All the ploughs were judged to be good on this difficult ground, but only the Howard and Ransomes ploughs considered good enough for further trials. Finally a dynamometer was obtained to resolve a dead-heat and the Ransomes & Sims plough was deemed slightly the better.

Such thorough and painstaking evaluation must have been invaluable to the plough makers, providing a rigorous performance comparison from which the future plough designs would benefit. Certainly for John Cooke it provided an endorsement of the design and construction of his wooden general-purpose plough which was still selling briskly in 1876, and it must have spurred his development of the patented composite beam which would greatly increase the strength of the plough.

By 1861 John Cooke was advertising 'Prize Ploughs' as his headline product and the following extract suggests that he had been actively involved in plough making from the mid-1840s. The fact that there were many small wheelwrights competing locally for the cart and wagon business possibly drew him into the more specialised plough making business where he could develop the smithy and iron foundry sides. He understandably continued the cart and wagon business and gradually introduced other farm implements which would sell readily to the same customers and help maintain a steady workload when orders for ploughs fluctuated.

> John Cooke has much pleasure in submitting this Catalogue to the notice of Agriculturists and the public generally, and embraces the opportunity to inform them that he has made the manufacture of **Ploughs** his **speciality** and **study** for upwards of **thirty years,** during which time he has gained such an amount of practical experience in this department as probably few men have, and which, he ventures to think, entitles him to the fullest confidence of his patrons. He also begs to announce that his Works are still under his own **personal supervision and direction,** and that his manufactures may therefore be thoroughly relied upon as regards **quality** and **efficiency,** and to properly perform the work for which they were intended; great **care** being taken as to the **excellence** of the **workmanship** and the **suitability** and **durability** of the **materials** employed.
>
> J.C. would draw the special attention of **merchants** and **shippers,** &c., to the fact that he possesses unusual facilities for manufacturing large **quantities** of **Ploughs** of any description on the **shortest notice,** having a large plant and staff almost solely engaged in making Ploughs.[3]

Fig 24
Ploughing matches at major agricultural shows demonstrated the qualities of ploughs and the skills of ploughmen; certificates were good advertisements for John Cooke & Sons.

Fig 25
Model XL RC was a light to medium-duty plough with Cooke's patented beam combining wrought iron and wood for strength and lightness.

Making High Quality Ploughs

A large proportion of the ploughs had John Cooke's patented construction of the beam using angle-iron and wood. The report in the *Implement and Machinery Review* of August 1878 states that this is a 'combination that not only lightens and cheapens the implement, but also gives to it an amount of strength otherwise unobtainable in many different forms'.

General Remarks, Description, &c.

The BEAMS are made of angle-iron and wood securely bolted and riveted together, by which means they are rendered exceedingly firm and strong, and impossible to be either strained or broken in work.

The WOOD HANDLES are made of best dry English ash or oak, simply put together with bolts, and are very easy to repair or replace; and the IRON HANDLES are very strongly constructed with diagonal stays, &c.

The MOULD-BOARDS or BREASTS and SHARES are on strict geometrical principles, and are the most improved pattern which a long practical experience has suggested, and for which Cooke's Ploughs are so celebrated.

The WHOLE OF THE PLOUGHS ARE MADE TO TEMPLATES ; therefore all parts of the same-sized Ploughs are exactly alike, and fit one another; so that if a part be broken or out of repair, an exact duplicate can at once be had from the Works, and any farm-labourer can replace the worn or damaged part in a few minutes.

All the CASTINGS, MOULD-BOARDS, SHARES, and other WEARING PARTS are of the best description and workmanship, manufactured only from the best quality of materials, and are so simply arranged that any labourer or boy may replace them.

The foregoing Ploughs are especially recommended to Merchants, Shippers, Colonists, &c., as they can be taken to pieces in a few minutes, and packed in a very small space for shipment. They may also be readily put together again when required, by any unskilled labourer, their construction being so simple; whilst their strength and durability, their suitability for all kinds of soil, together with the superior facilities they afford for repairs, renders them the most suitable Ploughs now manufactured for exportation.

An objection to Wood Ploughs - viz, the mortice holes which weaken the wood so much - is here entirely done away with, as there is not a single mortice hole in the whole plough, the beam being fitted into a metal socket which is screwed on to the hale, and the whole being put together with bolts [4].

That fact that parts were all produced to templates would clearly make servicing simple and emphasis was placed on the fact that no skill was required to replace parts or to assemble an implement after shipping. These would have been important considerations on remote farms.

Fig 26
Model LOT2 was a light plough capable of being worked by a pony on light soils and intended for small-holdings.

At the heart of the plough is the share. Two horses were used to draw a single-furrow plough in all but the lightest ground; a share which failed to remain sharp would slow the ploughing and exhaust the horses. John Cooke sold strongly on the attention he paid to high quality materials and consistent processing.

COOKE'S VERY SUPERIOR CHILLED SHARES

Always wear Sharp and never wear out of Hold.

It cannot be too strongly urged that, in order to ensure the right working of Ploughs, the shares should be made properly, and as there are parties now making shares said to fit Cooke's Ploughs, which they sell at a small reduction in price, in order to induce the purchasers to buy, but which are of a very inferior quality, being made of common metals, &c., and cause the Ploughs to run badly, John Cooke begs to call the attention of those using his Ploughs and others to his very superior and celebrated chilled shares, manufactured by a very superior process, and only from the best and most expensive quality of materials.

They are as hard as cast steel on the under surface and of ordinary hardness on the top, consequently **they never wear out of hold,** and **do not wear blunt,** as common shares do, but invariably wear off on the top surface only and to a **sharp edge right up to the socket.** Though nominally a little more in price, Cooke's shares are in reality much cheaper than the common shares, as they will wear two or three of them out, in addition to making the Ploughs to work so much better, especially in a dry season, when the inferior quality of shares, through wearing off on the under surface, will not face the hard ground at all.

N.B. - Chilled shares to fit nearly all the principal Ploughs now in use can be supplied on the shortest notice. Also shares made to any pattern when a quantity is ordered.

CAUTION - All Shares made by J. Cooke to fit his Ploughs are marked with his name, without which they are only imitations [5].

The following extract from the 1876 Catalogue shows that Cookes were aware of the export potential and prepared to adapt their ploughs to meet the needs of overseas customers:

> The Mark XL.ALL. No.2 combined Wood and Iron Plough adapted for deep and heavy land ploughing could be supplied with an extra long breast for Match ploughing. It could also be fitted with Kentish Breasts for turning the furrows. Version No 1 with iron handles is exceedingly strong and durable; very suitable for exportation; it has deep breasts especially for Russia. No.1 costs £5 11/-; No.2 costs £4 1/- [6].

Fig 27
This ⅓ scale model was used to demonstrate the construction, versatility and ease of servicing of Cooke's ploughs.

The illustrated catalogue of 1876 shows a view of the smiths' shop and forge at the Lindum Ploughworks *(Fig 28)*, having 3 bays each of about 9 m x 36 m with five hearths between the smiths and the forge, serving both departments, and a range of belt-driven machines between the smiths and an assembly bay. Light castings are being poured in the forge, anvils are being used in the smiths, and fitting work is taking place on benches and on the floor in the assembly bay. There is no assembly line, but the whole area looks active but uncluttered. It looks a practical facility for small-batch production of ploughs.

A fourth such bay was principally for wooden cart and wagon making, and wheel making. Stores were held in a separate building. At the Croft Street end was the steam-driven engine for the over-head drive belts. About 50 people were employed at the Lindum Ploughworks in 1876.

The Firm Gains Wide Recognition

John Cooke advertised and exhibited at every Lincolnshire County Show from 1869 onwards. *The Implement and Machinery Review* in August 1878 reported that Cookes showed twenty ploughs at the 1878 Royal Show and a year later exhibited 34 'splendid samples' at the Royal Show in Kilburn. On both occasions the *Review* wrote extensively and appreciatively about John Cooke's implements.

Ploughing matches provided an excellent opportunity to compare the performances of all the major plough makers' products; medals and awards were a strong selling point *(Fig 24)*.

In their 1876 catalogue Cookes drew attention to '20 First Prize Silver Cups and numerous other Prizes at the 23 most important Ploughing Matches in Lincolnshire and neighbouring Counties in the Autumn of 1874'. They also quoted the following press extract:

Great Trial of Double-Furrow Ploughs.

> At the Peterborough Ploughing Matches on Thursday last, there was very severe competition amongst the Plough Makers in the Double-Furrow Class. The competitors were Messrs. Ransomes & Sims, Messrs. J. & F. Howard, Messrs. Fowler & Co., Messrs. Ball & Son, Mr. J. Cooke, and Messrs. Hornsby and Sons, the latter firm entering two teams. The Society's Silver Medal and First Prize were awarded to Mr. J. Cooke, of Lincoln, whose Plough did splendid work [7].

Fig 28
An impression of the smiths' shop at Lindum Ploughworks showing the smiths' hearths flanked by the casting foundry and the assembly bay.

Clearly John Cooke's Ploughs had achieved a high reputation and were competing with five considerable companies. Hornsby of Grantham was their only local rival; the others were all located some distance away.

Medals and awards from major agricultural shows were prized achievements and would have been prominently displayed. John Cooke's products received many awards including: The Victoria Bronze Medal: Local Prize for Success in Art awarded by the Department of Science and Art, 1865 *(Fig 29)*; Charles, Earl of Yarborough Award, 1871; Lincolnshire Agriculture Society, Spalding Exhibition, President's Medal, 1872; Boston and District Agricultural Society Award for the Best Implement Adapted for Lifting Sugar Beet (undated, but after 1887) *(Fig 30)*; Wirral and Birkenhead Agricultural Society Award for the Best Collection of Ploughs, 1897 *(Fig 31)*; Staffordshire Agricultural Society Silver Award for the Best Collection of Ploughs, 1901.

The whole operation, as portrayed in John Cooke's illustrated catalogue of 1876, looks purposeful and well directed; the engineering was sound, the products well adapted and responsive to the market and the business effectively directed by John Cooke.

John Cooke's business had clearly been prospering during the buoyant agricultural period up to the 1870s. From small beginnings he had built a thriving business occupying new factory premises and then moving his household from Monks Road to a substantial residence on Newland Street West. 'Whitehall House' (also called Newland House) was a detached house with stables adjacent to what is now West Parade and with its own secluded park and carriageway to a lodge on Newland Street West. The property occupied about three acres and was surrounded by a high stone wall. Whitehall Grove was built on the grounds, and the house was demolished when Yarborough Road was created in the early 1900s.

Fig 29
Victoria Bronze Medal:
Local Prize for Success in Art awarded by
The Department of Science and Art
1865.

Fig 30
Boston & District
Agricultural Society Award for the Best Implement
for Lifting Sugar Beet.

Fig 31
Wirral & Birkenhead Agricultural Society
Award to Messrs. John Cooke & Sons for
Collection of Ploughs, 1897

John and Elizabeth Cooke had four children. Of their three sons, William and Frank were involved in the running of the company during the 1880s, but there is no mention of John, the third son, being active in the business. In 1885 John Cooke is listed as being a councillor of the City. In John Cooke's will, Elizabeth and daughter Mary were provided for by a trust on Wyves Farm, Fossdyke Bank, which had been purchased about 1880. John inherited the Plough Works premises; William and Frank shared the Plough Works business.

When John Cooke, the founder of the firm, died in 1887, the company took the name John Cooke & Sons and was managed by Frank, who was 26 years old.

There is no record of the further involvement of the other two brothers, who both appear to have left Lincoln at about that time. As there is no mention of the company in subsequent wills of William and John, it is possible that they both sold their shares to Frank.

After John Cooke's death, Frank Cooke and his family moved to 'Sunnybank' on Mill Lane overlooking the newly constructed Yarborough Road. By 1909 the Frank Cooke family had moved to 'Limefield' situated between Greetwell Road and Wragby Road with grounds extending to Wragby Road on the west and with a particularly fine view of the Cathedral. 'Sunnybank' was very much eclipsed by 'Limefield'; perhaps economies had to be made when taking on the Plough Works after John Cooke's death and the subsequent 20 years of business enabled an opulent residence to be bought.

The Twentieth Century: Less Successful Days

The buoyant agricultural economy in the early and mid-1800s had started to falter with the reduction of trade barriers after the repeal of the Corn Law in 1845 and crashed from the mid-1870s with the arrival of bulk imports of North American cereals. Although the livestock export trade continued well into the 1900s, the tide was starting to turn with the arrival of refrigerated meat shipments from the Dominions.

Many long-established agricultural machinery companies closed in the final quarter of the century; those which survived either had strong export markets or they diversified their products for other markets. John Cooke & Sons was rather exceptional in that, whilst it did broaden its export range, it still relied substantially on its home base and its established products.

Fig 32
The Turn-Wrest or One-Way plough
allowed the ploughman to save time by returning on the adjacent
row. The coulter-shares shown were less liable to bind or foul
in grass or stubble.

Fig 33
Frank Cooke took over the company on the death of his father John Cooke in 1887 and ran the business until his own death in 1922.

The Foreign and Colonial Catalogue of 1896 shows that John Cooke & Sons was continuing to develop its range of ploughs to include bean-sowing ploughs, turn-wrest or one-way ploughs *(Fig 32)*, a specialised plough for Bulgaria and vine ploughs for South America and Spain. The foreign catalogue also details colonial double-furrow and 3-furrow ploughs, ganged riding ploughs *(Fig 34)* and a full range of agricultural carts and wagons and sprung carts and drays.

Overseas sales were through local agents. Unlike the complex products made by many Lincoln companies which required a level of technical expertise for commissioning and repairs, Cookes' ploughs were simple, rugged and designed to be maintained by capable mechanics.

When Frank Cooke took over the firm there were about 70 employees; when he died there were about 100 employees. The medals and awards received after 1887 show that John Cooke & Son continued to promote and demonstrate their ploughs, and their continued operation on Frank Cooke's death in 1922 suggests that they continued to be competitive and successful. The obituary to Frank Cooke refers to the difficult trading period which had been experienced over the years prior to 1922; these difficulties were also probably experienced by their competitors in the depressed agricultural conditions following the end of the First World War. The fact that Frank Cooke's funeral was attended by a large number of employees, many of whom had worked there for over 30 years, was remarked on in the obituary as indicating the excellent industrial relations sustained by the company.

John Cooke & Sons' illustrated catalogue of 1934 *(Fig 35)* is recognisably a descendant of those for 1876 and 1896. It advertised: a single-furrow plough; 3 match ploughs; 14 general purpose ploughs; 7 one-horse/pony ploughs; 6 light ridging & earthing ploughs; 10 digging ploughs; 4 turn-wrest or one-way ploughs; 1 balanced plough; 8 double-furrow ploughs; 4 three-furrow ploughs. Among the plough accessories and attachments were shares with cast-in coulters; sub-soiler attachments; sower or drill attachments; lever ploughing-depth adjustment. Other implements advertised were 3 horse-drawn cultivators and sub-soilers; 9 horse-drawn hoes; 7 horse-drawn harrows; 2 horse-drawn rollers, flat or Cambridge; tractor ploughs; tractor harrows; tractor rollers; sack lifters, sack barrows; turnip cutters, turnip pulpers; cake breakers; potato sorters; pig feeders, pig troughs.

Fig 34
Cooke's Gang or Riding plough with lever draught adjustment, also available with coulter shares, was developed for uncleared land in overseas markets.

The final section showed agricultural carts with Dunlop pneumatic tyres; wagons; combined carts and wagons; drays or lurries; cattle floats and milk floats. The firm advertised that they could undertake general engineering work and casting work [8].

The list of ploughs had become much more extensive. Horse-drawn implements were almost universal into the 1920s, when tractors were starting to become more reliable, practical and affordable, and remained the dominant method well into the 1930s. Tractor-drawn implements were not listed in the general catalogue, but in a separate brochure.

The implements listed in the 1934 catalogue were probably nearly all developed and introduced from the mid-1870s onwards and defined John Cooke & Sons' product range into the 1920s. Many more plough types had been introduced, including additional one-way ploughs and drilling/sowing ploughs, but the majority of new products were small implements which could be sold to the same customers as the ploughs.

Fig 36
Cookes were still exploring new markets in the 1930s; a small tractor of unknown manufacture coupled to a Cooke roller photographed at the works.

These small implements were readily available from many manufacturers and so would have been keenly priced and consequently contributed little to the profitability of the firm. Such products would provide no additional defence to the business in times of agricultural recession.

The End of the Company

After Frank Cooke's death in 1922, the company was managed by his eldest son Sidney Cooke; the younger son, Horace, looked after the sales side. 'Limefield' was disposed of and Frank Cooke's widow Annie, daughter May and Horace moved to The Grove; Sidney and his family were living in Heighington.

Through the 1920s tractor-driven implements were developed and introduced and exhibited at all the major agricultural exhibitions and shows. Self-lift ploughs, one-way ploughs and three-furrow ploughs were all brought to market. Improvements were made to mould-boards to reduce friction by using a three-ply construction consisting of a sheet of soft centre iron enclosed between two sheets of nickel steel, the latter taking a very high polish. Although tractors were scarcely seen until the late 1930s, it was clearly necessary to be seen to be preparing products for the future.

The agricultural recession following the end of the First War continued during the 1920s and had become a part of the General Depression by the 1930s. The reference in the 1934 catalogue offering engineering and casting capacity indicates the difficult trading

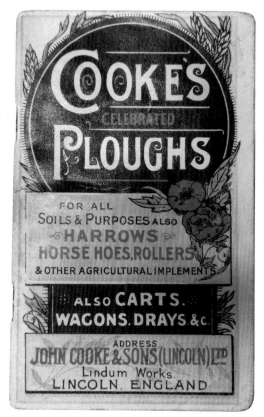

Fig 35
The highly decorated and brightly coloured covers of the 1934 brochure. The blue covers of the 1876 brochure were even more elaborately decorated.

conditions, as does the Reduced Price List introduced in July 1934. Evidence of a severe decline in business was the placement of small advertisements in the weekly Lincolnshire Chronicle in the early months of 1937 soliciting for wheelwrighting and general repair work, but no further advertisements were placed in the local press during the Lincolnshire Show or later that year. John Cooke & Sons was put into receivership in March 1938 with between 60 and 70 staff affected. In July 1938 the local press printed a brief history of the 80 years of John Cooke & Sons' Lindum Plough Works, reporting that 20 people were still employed and announcing that the assets of the business were to be auctioned in September 1938. (Some patterns, jigs and templates were acquired by J B Edlington of Gainsborough, see pp.62, 65).

It is evident that the company never removed the old implement types from its catalogue; the wide range of products must have been a nightmare to keep available. Preserving patterns and castings for designs which had been in the catalogue for 50 years must have been costly. There is no evidence that John Cooke & Sons significantly changed their manufacturing processes to reduce production costs; they remained a craft-based company to the end.

Would John Cooke, had he lived longer, have attempted to broaden his product range into new markets, as did many of Lincoln's engineering companies?

In its second generation, under Frank Cooke, the company continued to plough the furrow which had been so successful but probably became progressively less profitable in the agricultural recession which occurred after the First War. In the third generation, under John Cooke's grandsons, the company probably operated on slim margins and became less capable of breaking into new fields. It is a story not un-typical of family-run companies.

Notes

[1] John Cooke, illustrated catalogue, 1876
[2] Ibid
[3] Ibid
[4] Ibid
[5] Ibid
[6] Ibid
[7] *Peterborough & Huntingdonshire Standard,* 3 October 1874
[8] John Cooke & Sons, catalogue, 1934

Fig 37
Colin preparing Beauty and Prince to draw an XLSS match plough at Blyth, June 2007.

CHAPTER 3

JAMES COULTAS OF GRANTHAM

PRIZE DRILL MAKER

Catherine Wilson

The Family

Machine maker, iron and brass founder, engineer, millwright, agricultural implement maker, agricultural engineer - so is the firm of James Coultas variously described in the trade directories between 1820 and 1900. But during the second half of the nineteenth century the firm's most notable successes were the 'First Prize Drills Patronised by Her Majesty the Queen' and the 'only perfect manure distributor' *(Fig. 39)*. How did the firm develop and why did they choose these specialisms? It is an intriguing story of enterprise and ingenuity, of dedication and determination, of a working class Methodist family becoming Church of England and 'establishment', and of achievement and family rivalry.

The first James Coultas (1788-1862) *(Fig. 38)* was born at North Burton (or Burton Fleming), near Bridlington, Yorkshire in 1788, the fourth child of a large family. There is a family tradition that he was the son of a velvet maker[1], but the family may also have been involved with blacksmithing.[2] He moved to Grantham in 1808 and set himself up in business. The earliest definite reference to him appeared in an advertisement in the Stamford Mercury of 10 May 1811, where he described himself as a 'threshing machine maker from Hull' and offered 'an improved portable machine for sale or hire'. He later became an iron worker and machine maker in Union Street, Little Gonerby on the northern edge of Grantham.

*Fig 38
James Coultas I, 1788-1862.*

*Fig 39
Advertising board, late 19th century.*

James's reason for moving to Grantham is not known, but he soon settled down and married a local girl, Miss Mary Trolley of South Stoke, in 1815.³ They set up home at 51 Union Street, close to the works, and had four children: James, born in 1819; a girl, Mary, thought to be James's twin, who died in infancy; Ann, born in 1825; and John, born 1826. James must have been a man of enterprise to have bought land and launched his own business by the age of 26. It is interesting to compare his progress with that of Richard Hornsby, who started at much the same time. Both men were Methodists⁴ and were both agricultural engineers and machine makers, but by 1850 Hornsby had built up an international business, whilst Coultas was still operating in a primarily local market, and there are other indications that he perhaps did not live up to early promise.

The second son, John, joined his father in the Little Gonerby business and is described as a machine maker in the 1851 census, when he was still living at home with his widowed father. By 1861 he had set up home in Union Street, close to the works, with his wife Louisa, daughter Ann and infant son John William. At this time John is described as an agricultural implement maker, brass and iron founder and millwright, employing ten men and six boys.⁵

James Coultas senior died in June 1862, and in that same year there was a fire at the Little Gonerby Works and considerable damage was caused. After 1862 the Little Gonerby works ceased to be mentioned in the trade directories and John Coultas's family is not mentioned as living in Grantham in the 1871 census.

But the main subject of this article is James junior (1819-1890), hereafter referred to as James II. He clearly inherited the engineering skill of his father, and it was he who went on to establish a national and international reputation, particularly for seed drills and manure distributors. However, for some reason, he did not join his father in his business as might have been expected, and as his brother John did, but set up on his own.

James II married Miss Elizabeth Perkins of Holbeach in 1840, and apparently went to live in Holbeach at that time. It was not unusual for the 'heir apparent' to a family business to work for other firms under an apprenticeship and to gain wider experience. But James II is recorded as 'working on his own account' in Wisbech where he 'succeeded in forming a lucrative trade'.⁶ When he returned to Grantham in 1852/3, he did not join the family firm, but started an agricultural engineering business which was 'altogether independent of his father's business, with which he had no connection whatever'. This is confirmed by the obituary in the *Implement & Machinery Review* which states that 'he had to compete against his own father' when he first set up his works on Wharf Road on the southern edge of Grantham in 1852/3.⁷ So there seems to have been a rift between father and son. Perhaps James senior was not as ambitious as James II, though we will never know why the latter chose confrontation with his family by moving back to Grantham, rather than continuing to work elsewhere. In the 1861 census he was described as an engineer employing 15 men and 22 boys. By 1871 he was an 'agricultural engineer and master' employing 48 men and 10 boys.

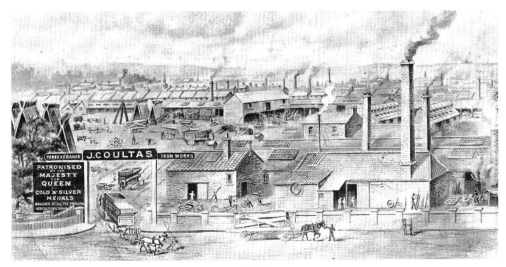

Fig 40
Back cover of catalogue about 1892, showing Perseverance Iron Works from Station Road.

James II was clearly inventive and entrepreneurial: 'Quick to perceive the value to an implement firm of the awards and commendations of the Royal Agricultural Society of England, Mr. Coultas seems to have set himself determinedly to the task of occupying the top place in drills, at the yearly exhibitions of that important body' He was also 'a most energetic businessman. Nothing upon which he had set his heart was too hard for accomplishment'.[8] From the obituary in the *Grantham Journal*, we learn that 'Mr. Coultas was an early riser and a most diligent worker. Thoroughness and perseverance were two of his distinguishing characteristics'.

He also undertook a variety of public duties, filling 'every parochial office in Spittlegate', before that parish became part of the borough of Grantham. 'He declined to be a member of the town council, but was a good churchman and a Conservative'. Even allowing for the fulsomeness found in obituaries, we have a picture of a hardworking, ambitious businessman, well respected and with a social conscience - perhaps typical of his age, and of the leaders of other such firms in the county. Certainly under his leadership the firm developed a national reputation and exhibited internationally, though still on a small scale compared to his neighbour Richard Hornsby. 'Seest thou a man diligent in business; he shall stand before kings' was considered a fitting epitaph for James II by the national journal, the *Implement and Machinery Review*.[9]

Fig 41
James Perkins Coultas, 1845-1910.

James II and Elizabeth had eight children: six boys and two girls. His four eldest children were born at Downham Market, Norfolk, before his return to Grantham. Of these children, the two eldest boys, James Perkins (James III) and Thomas, joined their father in his flourishing business and took it on after his death. George died young, whilst William, Benjamin and John Vickerman were involved in another successful business in Grantham as general ironmongers and timber merchants. The two girls, Elizabeth and Briseis, both married, left Grantham, and were not involved in the business.

His eldest son, James Perkins (1845-1910) *(Fig. 41)*, seems to have been of similar character to his father.

He was articled to William Foster of Lincoln, where he 'acquired a thoroughly practical acquaintance with the details of his profession'.[10] He then worked for a time on the Great Northern Railway at Peterborough before joining his father in the business. With his father, he is credited with many of the improvements to the seed drills and other machinery; indeed there are patents in his name. After his father's death in 1890, he carried on the business very successfully.

When he was elected Mayor of Grantham in 1897, the *Implement and Machinery Review* carried a detailed article about him. The article described him as 'a very pleasant man to speak to' and being 'of a cheery and apparently hopeful disposition, he is just the man to secure the best possible reception for the implements and machines which he manufactures'. He was also 'thoroughly practical and knows exactly what he is talking about as an agricultural engineer'.[11] That was high praise indeed from the national journal. Thirteen years later, the same journal carried a lengthy obituary.[12]

Amongst his public offices, he was an Alderman for Grantham, and a member of Kesteven County Council. He was a sidesman and then Vicar's warden at St. John's Church, Spittlegate. And, like his father, he was a Conservative 'though he took no active part in electioneering'. The *IMR* reporter commented that he was 'not as vivacious as some of our friends' but was 'very thoughtful'. He was: 'a man of good sense and one who has very materially helped to build up and sustain the prestige of the British agricultural drill-making industry'

This is not the appropriate place to detail the careers of all James II's children, but brief mention should be made. Thomas, the second son, joined his father and brother in the family business and was an integral part of its success. In 1905 he decided to retire and asked James Perkins to buy out his share of the business. Despite the strain this put on the finances of the firm, this was done. Thomas continued to be active in the local community until his death in 1915. His son, Thomas Bestwick Coultas, was not involved in the family business and was killed in the First World War.[13]

The three younger sons of James II, William, Benjamin and John Vickerman, were involved in a subsidiary business as general ironmongers and timber merchants. William took over premises at 91-92 Westgate and set up as an ironmonger, also selling Coultas's products. He died at the age of 28, but his two brothers took it on. In 1882 the business was described as 'general and furnishing iron merchant, iron monger, bell-hanger and gas fitter'.[14] By 1889 the description was even longer: 'general and furnishing ironmonger and cutler, marble chimney pieces, register stoves, grates and kitchen ranges, agricultural implements, iron, steel, file and tinplate merchants'.[15] Ben and John Vickerman continued to run a successful business into the 1920s. The timber merchant part was situated on Wharf Road, just to the north of the Perseverance Iron Works *(Fig 44)*.

Fig 42
James Riley Coultas, 1880-1952.

James Perkins (James III) married Esther Riley and they had two children: a son, another James, known as James Riley, and a daughter Catherine. James Riley Coultas (1880-1952) *(Fig 42)* joined his father in the family business and took over on his father's death in 1910. He managed it through the difficult years of the First World War until 1920, when it became a Limited Company. Hornsby had also had a difficult time up to and during the war, and amalgamated with Ruston, Proctor & Co of Lincoln to become Ruston & Hornsby in 1918. The Coultas business was sold to David Roberts, joint Managing Director of Ruston & Hornsby and previously Managing Director of Hornsby, 'at a loss'.[16]

In 1905 James Riley Coultas married Margaret Pitchford, a young Australian woman, and in 1925 he emigrated to Australia with his wife and their three young children.[17] He died in 1952, thus ending the direct connection between the Coultas family and the business. James Coultas Ltd, however, carried on trading until it was finally closed in 1955. At that time John H Rundle of New Bolingbroke purchased the patterns and spares for the Coultas drills and rising hopper distributors, and continued to manufacture spares until 1970.

Little Gonerby and The Perseverance Iron Works

From 1808 until 1814 the first James Coultas rented a piece of land that he was then able to buy. It was in Little Gonerby close to the Great North Road and had a good water supply from the Mowbeck.

A deed, dated 14 June 1814, survives at Lincolnshire Archives recording the sale of this piece of land, 160 square yards in extent, by Joseph Tindal of Manthorpe, tanner, and Robert Cox of Grantham, hatter, to James Coultas of Grantham, machine maker, and Thomas Lyne of Grantham, ironmonger. The cost of this was £140. A small plan attached to the deed shows the North Road on one side, an 'intended street' on a second side, and private owners on the other two sides.[18] The 'intended street' became Union Street, which was the address for both the works and the home of James Coultas. *(Fig. 43)*

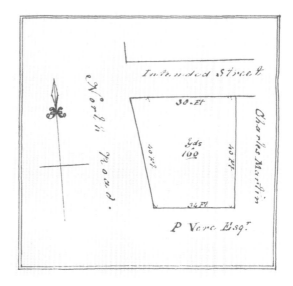

Fig 43
Plan of land acquired by James Coultas in 1814.

The works at Little Gonerby are recorded in directories from 1826, with Coultas listed as a 'machine maker'. The 1855 *Post Office Directory* listed James Coultas & Son as 'machine makers, millwrights, iron & brass founders'. The range of products was considerable and included horse hoes and other agricultural implements, but with seed drills as a speciality.[19]

Sadly, nothing further has been discovered about the layout of the site. The buildings have been completely demolished and the site is now the car park for a large supermarket. No photographs, illustrations or actual examples of the products are known to survive.

James II set up his rival business initially on a site on Wharf Road, though no trace now remains. Then in 1862 he acquired land alongside the Great Northern Railway. The site was a triangle between Queen Street and Station Road with a footpath, Station Path, marking the northern boundary of the site *(Fig 44)*. At the southern point of the triangle Launder Terrace runs at a right angle to Station Road, and it was here that the Coultas family home, Home House, was situated.

Fig 45
General view of workshops, from 1892 catalogue.

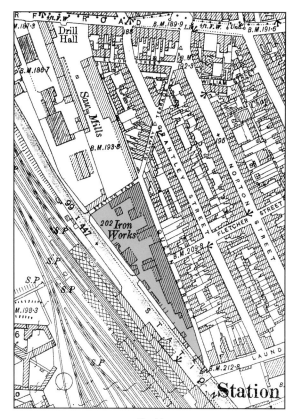

Fig 44
1906 edition Ordnance Survey map showing location of Perseverance Iron Works between the Railway Station and Wharf Road.

The house has now been demolished, and there are modern properties on the site. Coultas named his new premises the Perseverance Iron Works, proclaimed on a large sign over the entrance *(Fig 40)*. Beyond this to the north, adjacent to Wharf Road, were the sawmill and timber yard of the other Coultas brothers, Benjamin and John Vickerman.

It has not been possible to trace the development of the site in any detail. The buildings must have been extended and developed as the business grew, but no detailed plans have been found before the 25in OS map of 1906, and no illustrations until the very end of the nineteenth century. The *IMR* obituary for James Perkins, previously cited, states that after his father's death in 1890, James P made the business 'hum, frequent extensions of the works being necessary'.[20] A catalogue from c1892 is the only one to have both an illustration of the works and actual interior and exterior photographs. The illustration is a 'bird's eye view' *(Fig 40)* and, as was common with such drawings, the perspective used tends to exaggerate the size of the site. However, close comparison of this view with the photographs shows that at least the main buildings are accurately shown, so we can probably assume that not too much artistic licence has been used.

The entrance was roughly in the centre of the east side of the site, off Station Road. There was a large sign above the entrance and a notice proclaiming the patronage of Her Majesty the Queen, so that no-one arriving at Grantham Railway Station could be unaware of the importance of the firm. The foundry was on the east side nearest to Station Road, with the moulding shop behind, running north-south. A parallel north-south building was probably the general workshops. Behind that again was a house, perhaps the foreman's house, and then an open yard which appears to have been used for assembly and some storage. Against the western boundary of the site, but at right angles to it, was a timber-clad building with open bays at ground floor level, and possibly the carpenters' shop above. To the north of this, the wider end of the site appears to have been used primarily for timber storage.

The foundry buildings were of brick and pantiles, whilst the other buildings look as though they had slate roofs. The site does not show any overall design and has clearly developed over time, with new buildings being added as needed.

No trace of the buildings remains. The site is used as the Station car park and all that can be seen of the Perseverance Iron Works is part of the brick boundary wall, together with some cobbles which are just visible underneath the tarmac of the car park.

Seed Drills

The photographs in the 1892 catalogue referred to above show some of the range of products that were being made at the end of the nineteenth century. Portable steam engines, threshing machines, elevators, hay rakes, reapers and carts all appear. However, it seems likely that James II concentrated primarily on producing first class seed drills, for which he won many prizes in the 1860s and 1870s, and then expanded the range of products as the business developed and as his son, James Perkins, began to play a larger role. As the seed drills are by far the best known of his products, they deserve detailed attention here.

Agricultural technology was moving rapidly in the middle of the nineteenth century and many new machines were being invented and developed to make life easier for the farmer. In 1851 Philip Pusey, writing in the *Journal of the Royal Agricultural Society for England* (RASE), commented: 'The sower with his seedlip has almost vanished from southern England, driven out by a complicated machine, the drill, depositing the seed in rows, and drawn by several horses.'[21]

In that year Richard Hornsby received an award for his corn and seed drill at the Royal Show, but there is no mention of the name Coultas. The earliest mention of James II gaining prizes for his drills appears to be in 1856, when he exhibited and won prizes at the Yorkshire and Lincolnshire shows.

The first national success was at the Leeds meeting of the RASE in 1861 (the Royal Show). Ten drills were selected for trial belonging to Mr James Coultas junior, Messrs. Coultas and Son, and six other manufacturers. Careful experiments were made to test the quantity of seed delivered per acre. Accuracy of delivery, construction and selling price determined the award. In the ten classes tested, James II won five first prizes and two Highly Commended awards. In the two classes for General Purpose drills, his father and brother won second prizes, but nothing in the other classes. We can perhaps imagine what the atmosphere was like between them on this occasion! Without wishing to overplay the possible family feud, it does seem that James II deliberately set out to produce drills better than those of his father. Interestingly though, Hornsby does not seem to have entered the competition on this occasion.[22]

The next major exhibition James II attended was the International Exhibition in London in 1862. Here, his exhibit received an Honourable Mention and a number of items 'were purchased for Her Majesty', presumably for use on the Royal farms, and this allowed James II to tell the world that his products were 'Patronised by Her Majesty the Queen'.

At the Plymouth meeting of the RASE in 1865 there was another trial of drills and manure distributors, and on this occasion Hornsby was competing, as well as other manufacturers. There were eleven classes for drills. James II received prizes or commendations in eight of these, including a First Prize for his hillside

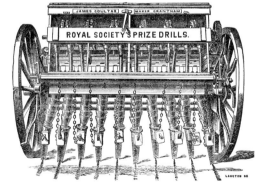

Fig 46
General purpose drill awarded First Prize at the RASE Bedford trials in 1874, from 1875 catalogue.

delivery drill. Hornsby also achieved one First Prize for a small seed drill, with awards and commendations in five other classes. The top honours for drills that year were, however, taken by John Sainty of Burnham, Norfolk.[23]

1874 was the year of Coultas's greatest triumph at a RASE event. Extensive trials of implements took place at Bedford, lasting an entire week. 'Probably no previous trials by this Society have presented more points of interest, or are calculated to prove more useful and instructive both to exhibitors and to the agricultural public, than the present'.[24]

There were ten classes for different types of drill. The drills were particularly tested for their regularity of delivery, and the trials were exhaustive. To test for regular delivery 'small bags were hung on the upper seed tin of each coulter, and after the run their contents were carefully weighed by the assistant-engineer.' The report of the trials gives considerable detail about each drill and its performance, and makes interesting reading, but can only be summarised here.

For general purpose drills, Coultas took First Prize, with Thomas Harrison of Burton Road, Lincoln in second place *(Figs 46 and 186, p152)*. For corn drills, Coultas again took First Prize, with Holmes & Son of Norwich in second place. 'The workmanship and finish of this drill are very good, and its price, £27, is moderate.' Out of a possible 1,000 points, the Coultas drill received 934, 130 more than its nearest rival.

In the other seven classes, Coultas took four First Prizes, one Second Prize and one Highly Commended. The drills for which he won prizes were: General Purpose Drill; Corn and Hill-side Drill; Turnip and Manure Drill; Corn and Seed Drill; Turnip, Manure and Liquid Manure Drill; Small Seed Drill; Small Occupation General Purpose Drill; Small Occupation Corn Drill; and Manure Distributor. No other company

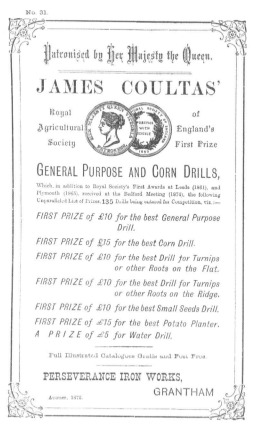

Fig 47
List of prizes in the front of the 1875 catalogue.

came near to rivalling this, though it was noted that exhibitors who had taken the top honours at the Plymouth trial in 1865 were not competing on this occasion. It was nevertheless an impressive achievement.[25]

It seems that Coultas had developed a method of seed delivery that was more accurate and reliable than his rivals, and his reputation was built on this and on the fact that the drills were robust and well-built. In addition steam drills (pulled by cable between two steam engines) 'suitable for every system of Steam Cultivators' were made to order.[26]

Coultas lost no time in using this success as a marketing tool, listing all the prizes at the front of his 1875 catalogue *(Fig 47)*, and subsequent catalogues. He also had successes at international exhibitions - Hamburg 1863, Paris 1867, Vienna 1873 and Paris again in 1878 - as well as numerous local awards.

From the 1860s until the death of James Perkins in 1910, Coultas seed drills and other products were shown every year at the Royal Show in July and the

Fig 48
The yard looking south, from 1892 catalogue.

Smithfield Show in December. The monthly journal *Implement and Machinery Review,* first published in 1875, each year carried reports of these two annual shows and the machinery exhibited there. Coultas features nearly every year, so there is a wealth of material, and the descriptions are often accompanied by illustrations. A few highlights are as follows:

Aug. 3, 1880 - Amongst the items on display was a two-row turnip ridge drill and a corn drill 'with force feed apparatus similar to those of American manufacture'.

July 1, 1892 - 'The largest and best collection of corn and seed drills at Warwick [Royal Show] was contributed by Mr. James Coultas'. There were four new drills, two of which were illustrated.

Jan. 1, 1893 - Coultas's 'improved folding elevator and stacker' is mentioned. Several were shipped to Spain for elevating cork.

June 2, 1896 - Coultas's potato planter is illustrated. It was a two-horse machine capable of planting ten acres a day.

Feb. 2, 1898 - Mr. Coultas was in discussion with a Chinese gentleman about using his drills to sow rice!

Jan. 2, 1901 - A new artificial manure drill was illustrated. 'The brothers (James Perkins and Thomas) were present, each of them looking as happy as ever.'

July 2, 1904 - There was a picture of the Coultas stand at the Royal Show, with James Perkins in his bowler hat and tail coat leaning proudly on one of his machines *[Fig 50]*.

Jan. 1, 1909 - 'The largest stand in the Hall, devoted exclusively to drills and manure distributors, was that occupied by Mr. Jas. Coultas, one of the oldest Smithfield Show exhibitors'. It was noted that a corn drill had recently been sold to India.

Fig 49
Portable engine with J P Coultas patent chimney lifting device.

In 1907 the Royal Show was held on the West Common in Lincoln. As might be expected, Coultas put on a good show, with no fewer than thirteen different drills. He also showed: four manure distributors, two horse hoes, an elevator and straw stacker, horse gear, Cambridge roll, flat roll and turnip cutter, all made by the firm. In addition, the display included ten items made by other companies, including a Hornsby binder, a Gratton of Boston sack lifter and harrows by W Ashton and Son of Horncastle.

Other Coultas Products

Although the name of Coultas is primarily associated with seed drills, the firm did build other machinery and could supply most farm implements and machines. This area of their work seems to have developed from

Fig 50
The Coultas stand at the Royal Show, 1904, with James Perkins Coultas in the bowler hat.

Fig 51
Elevator and horse gear, c1879

the 1870s and diminished after 1900, but the 1879 catalogue illustrates: a portable steam engine (James P Coultas patent), which could be supplied from 4 hp to 12 hp, ranging in price from £150 to £280; a thrashing and finishing machine in sizes of 4 ft, 4 ft 6 in., or 5 ft at prices from £125 to £160; an elevator for hay, corn or straw, costing £50 *(Fig 51)*; a cast-iron horse gear with covered safety dome top for one or two horses; a winnowing machine; an oil-cake crusher; and a Cambridge roll. The invention by James Perkins

Fig 52
Rainwater gully, Wide Westgate, Grantham.

about this time of an apparatus for raising and lowering the chimney of a portable engine was acclaimed as a simple and effective addition to any engine *(Fig 49)*. 'In connection with boilers, we may mention a very simple device shown by Mr. James Coultas, of Grantham. A handle on the smoke box works an endless screw, which turns a small windlass, and thereby winds up the chimney'.[27]

In addition to agricultural machinery, the firm produced a lot of general cast-iron work such as kitchen ranges and street furniture - manhole covers, rain water gullies *(Fig 52)* and the like.

Several of these can still be seen on the streets of Grantham, though sadly they do not seem to be recognised as significant by those who maintain these streets. There are also a number of cast-iron street signs in the immediate vicinity of the Perseverance Iron Works *(Fig 53)*. These have a diamond motif, similar to that found on the rain water gullies which bear Coultas's name. It seems reasonable to assume that these street name-signs too were cast at the Perseverance Works.

What else survives? There are two Coultas drills in the Museum of Lincolnshire Life and a few remain in private hands in Lincolnshire and elsewhere. Grantham Museum has an interesting collection of printing blocks, clearly used for printing the firm's various catalogues. Some cast-iron finger posts in Kesteven were made by Coultas; whilst a cast-iron pillar and sack-hoist mechanism survive in the barn at Home Farm, Stoke Rochford. (Information on the whereabouts of any further Coultas products would be most welcome).

After 1910

The death of James Perkins Coultas in 1910 must have had a significant effect on the family firm, but it continued to exhibit at the shows, particularly the Royal, displaying a range of seed drills and other equipment. At the Smithfield Show in December 1919 it displayed a small-seed and rye grass drill, suitable for sowing linseed; a corn drill and four manure distributors.[28] Developments were clearly taking place, since the *IMR* for February 1920 reported that the firm had produced an adaptation of their standard drill for use with a tractor, and illustrated the arrangement.[29]

At the beginning of 1921 the firm became a limited company known as Messrs James Coultas Ltd. It continued exhibiting at the shows and reports demonstrate that the product range was continuing to develop and change. For example, a new 'R.O.' drill was introduced in 1927; a 'Discodrill', a convertible drill or hoe, in 1928; and in 1930 their manure distributors were 'meeting with an ever increasing sale'.[30] Some works photographs survive from around this time, showing that the basic design of seed drill had changed little, though they were adapted for tractor haulage *(Fig 54)*.

The 1930s were a busy time with several new products introduced, particularly a range aimed at smaller scale use for market gardens, top dressing lawns, tennis

Fig 53
Cast-iron street name, Grantley Street, Grantham.

courts and sports grounds. These were no doubt designed to offset to some degree the effects of the depression in the general agricultural market. Catalogues and price lists were still being produced, though with a more limited range than in earlier periods. A leaflet describing 'The Coultas Fertilizer Distributors' was produced in 1950. Amongst users listed were the Royal and Ancient Golf Club, St Andrews, and the All England Lawn Tennis Club, Wimbledon.

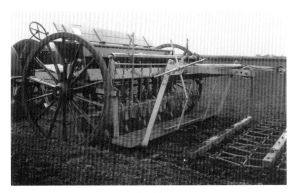

Fig 54
Tractor-drawn seed drill with attached harrows.

However, the market for their traditional agricultural product was rapidly disappearing and in 1955 the Perseverance Iron Works closed and the property was sold. All the contents were sold at auction on 23 November 1955, including the tools from the blacksmiths shop, wooden benches, 'three screw bottle jacks', various types of paint, 'useful timber' and 'various office furniture and fittings'.[31]

But that was not quite the end, as John H Rundle, Ltd of New Bolingbroke purchased the patterns and spares for the Coultas drills and continued manufacturing spares until 1970, when a few parts were still in stock.[32]

So came to an end a family business which had not only developed an excellent national reputation for that essential piece of farm equipment, the humble seed drill, but also played a significant part in the local community for 150 years. But just what was the relationship with Hornsby through the years, from the time of the Methodist 'founding fathers', through the period of prosperity in the 1870s and 1880s when both their factories and their homes were more or less next door to one another, to the 'take over' in 1920?
It is interesting to speculate.

Notes

[1] M J Lamyman [now Whittington], *Stand before Kings*, unpublished manuscript, 1992, p.1
[2] Ann King, *pers. com.*, 2005
[3] Lurleen Slaney, *pers. com.*, 2006
[4] Michael Pointer, *Bygone Grantham* (1978), p.23
[5] The ten-yearly census returns from 1841-1901 have been consulted for this article
[6] Obituary, *Grantham Journal*, 1890
[7] Obituary, *Implement and Machinery Review*, 1 Nov 1890
[8] *IMR*, 1 Nov 1890
[9] *Ibid.*
[10] *IMR*, 1 Nov 1910, p.851
[11] *IMR*, 2 Dec 1897, p.22246
[12] *IMR*, 1 Nov 1910, p.851
[13] Lamyman, *Stand before Kings*
[14] W. White, *Directory of Lincolnshire*, 1882
[15] *Kelly's Directory of Lincolnshire*, 1889
[16] Lurleen Slaney, *pers. com.*, 2006
[17] Lurleen Slaney, *pers. com.*, 2006
[18] Lincolnshire Archives Office, BRA 1487/10/2
[19] Lamyman, *op.cit.*, p.2
[20] *IMR*, 1 Nov 1910, p.851
[21] *Journal of the Royal Agricultural Society of England*, Vol. 12, 1851, p.600
[22] *JRASE*, Vol. 22, 1861, p.454
[23] *JRASE*, Vol. 26, 1865, p.376-383
[24] *JRASE*, Vol. 35, 1874, p.627
[25] *JRASE*, Vol. 35, 1874, pp.626-662
[26] James Coultas, *Illustrated Catalogue of Royal Society's First Prize Drills and Agricultural Implements*, 1879
[27] *The Engineer*, 20 July 1883
[28] *IMR*, 1 Jan 1920
[29] *IMR*, 1 Feb 1920
[30] *IMR*, 1 July 1930
[31] Sale brochure, 23 November 1955, in ownership of Alan Rundle
[32] Letter from J H Rundle to David King, 10 April 1973, in ownership of Alan Rundle

CHAPTER 4

RICHARD DUCKERING OF LINCOLN
IRONFOUNDER

Mark Duckering

The Origins: Burton and Duckering

Richard Duckering was born on 28 March 1814 at Coates near Stow, the son of Samuel Duckering, a farm servant, and one of nine children. A staunch Primitive Methodist, living on Waterside South in Lincoln at the time of the 1841 census, he was married with a son and heir, Charles, a month old, and his occupation was given as iron master.

By 1845 Richard had gone into business with Edward Burton to form the partnership Burton and Duckering, ironfounders, which was one of the earliest in the city. Both employers worked alongside their employees; Duckering was a moulder, Burton a machine maker.[1] Their premises were on Waterside South at number 53½. 'There were the foundries of Ruston and Proctor, of Burton and Duckering and of Clayton and Shuttleworth reached in turn', according to a contemporary record.[2] Examination of the Padley plans of the city for 1851, 1868 and 1883 suggests that the foundry site was close by the southern end of the present footbridge across the Witham, in line with Montague Street.[3]

In 1847 disaster struck when Burton and Duckering's foundry suffered a serious fire. The *Lincoln Rutland and Stamford Mercury* of Friday 3 September 1847 reported the event thus:

FIRE

A few minutes before 11 o'clock on Monday night the inhabitants of Waterside were alarmed by perceiving the foundry of Messrs Burton and Duckering burst all at once in a terrific blaze which enveloped the whole building. The brilliant mass of flame in the course of a very few minutes attracted a large expanse of people and the County and City engines were speedily upon the spot, but the fire raged with such fearful intent that all attempts to go inside the building and rescue from destruction any of the valuable models were utterly futile. So excessive and general was the fire that to hazard human life within its reach would have been madly courting disaster.

The power of the engines however were brought speedily upon the burning mass and in about half an hour the fire was subdued beyond all danger.

The building was composed nearly entirely of wood and never ought to have been used for so dangerous a business as smelting and casting iron.

Many of the workmen had not left their places more than half an hour before the conflagration occurred. All then appeared well but it is quite evident that the slightest particle of fire would rapidly extend amongst timbers so dried by their constant exposure to heat that they would blaze like match wood.

The flames appeared to burst from the whole building at once; this is so counted for by the fact that the crevices of the roof were filled with soot. Until the fire in the interior acquired sufficient power to force this off there would be little perceptible blaze; this soot seemed at length to take fire like gunpowder and erupted in a magnificent coruscation and the fire in the interior, having then obtained an upward vent, raged like a furnace.

All the valuable models, many of them belonging to different persons in the County, were completely destroyed.

The total damage is estimated at £500 and neither the buildings nor contents were insured.[4]

Charles Duckering (his son) in 1912 described the events and cause of the fire as follows: 'While carrying out certain work in the foundry one day, the liquid hot metal squirted up into the roof and dropping back down again with the accumulations it brought from there fell into the blacking and oily matter, creating an instant blaze. Almost as quickly as can be told the flames leaped up and soon the dry timbers were ablaze. The fire brigade were too late arriving and when the fire was put out there was very little left.'[5]

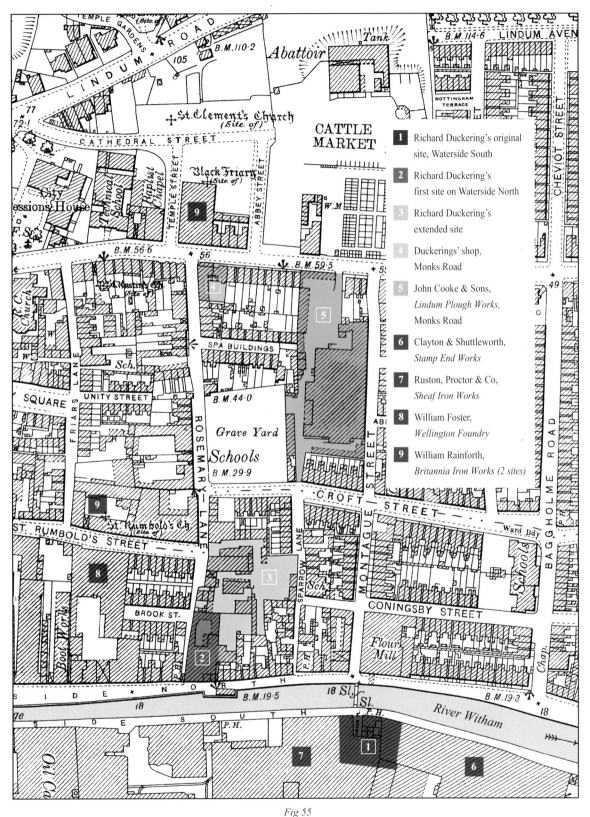

Fig 55
Map of Waterside and Monks Road area of Lincoln, 1905.

Richard Duckering was 'a man who did not sit down amid the ruins and mourn over his losses. As soon as they could rig up tarpaulins and get something like a roof across the temporary timbers work began again'.[6] The following year the firm was advertised as manufacturers of ovens, boilers, plough, drill and threshing machines castings.[7] Burton and Duckering parted company some time between 1850 and 1854.

In 1856 Charles Duckering had completed his schooling and so joined his father's company full time at the age of fifteen. He had already helped his father while still a young boy by addressing labels for the deliveries which were made to the steam packet companies on the wharf side. 'Taking orders at home after school and later, his evenings were spent at the foundry in the way of book-keeping and so on. He now went into the firm in the practical all-round fashion of the day doing clerking, moulding and a variety of other work'.[8] The works grew and they employed between 25 and 30 people at this time. The factory was now too small and so a move was made to a new and larger site on the northern side of the River Witham.[9]

The New Site on Waterside North

The mortgage for the land Duckering chose appears to have been taken out by Richard's brother, Thomas Duckering, a gentleman's servant, living at Ashby-de-la-Launde. Edward Ross, a farmer from South Carlton, Nottinghamshire, put up '£640 plus interest of five pounds per centum per annum'. This site was a matter of 100-150 metres from their previous home. (The location of the main workshop was where the YMCA building stands in 2006 and the yard is Siemens' car park) *(Fig 55)*. This new site was described in the article of agreement as 'land, gardens and outbuildings belonging to William Kirk and John Hind together with coal yard and buildings lately occupied as offices by The Witham Steam Packet Company. And also all that plot of ground …at the back containing some 420 square yards.'[10]

The purchase money had to be paid by 15 May 1857 for the deal to be finalised. The move was made between 1857 and 1859 and 'was one of the first outward and visible signs of the reward met with by the enterprise of the founders of the firm. The business spread its reputation in a manner very prompt for so young a firm, and the highest credit must be accorded the keen business ability and untiring energy of Duckering's in consequence'.[11]

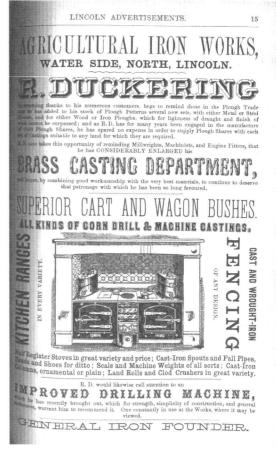

Fig 56
Advertisement in Morris & Co's Commercial Directory and Gazetteer of Lincolnshire, 1863.

Little is known of the firm's early years on Waterside North, but an advertisement from around this time (c1863) gives an insight into what was available from Richard Duckering's Agricultural Iron Works: 'Superior cart and wagon bushes, all kinds of corn drill and machine castings, cast and wrought iron fencing of any design. Richard Duckering also calls attention to his recently brought out "Improved Drilling Machine" which for strength, simplicity of construction and general usefulness warrant him to recommend it. One is constantly in use at the works where it may be viewed.' He also called upon those in the plough trade with his 'stock of new sets of plough patterns with either metal or steel breasts made for either wood or iron ploughs which for lightness of draught and finish of work cannot be surpassed.' This advertisement also stated that 'he has been a manufacturer of chill plough shares for many years and has spared no expense in order to supply plough shares with each set of castings suitable for any land for which they are required'.

Millwrights, machinists and engine fitters were not left out either: 'His considerably enlarged Brass Casting Department is brought to their attention.' There are: 'kitchen ranges in every variety, half register stoves in great variety and price, cast iron spouts and fall pipes, head and shoes for them too; scale and machine weights of all sorts; cast-iron columns, ornamental or plain, land rolls, and clod crushers in great variety.' *(Fig 56)*.[12]

Charles Duckering Inherits the Foundry

On 23 August 1870 Richard Duckering died, aged 56 years, and his only son Charles inherited the foundry, giving it his name soon afterwards. Charles, his wife Sarah Ann and two daughters lived at 26 Monks Road, opposite Rosemary Lane, a stone's throw from the business, which continued to expand.

Another advertisement, this time from 1877, shows that they continued to manufacture kitchen ranges, register sham and mantel stoves, plough castings, steel breasts and plough shares, girders, columns and palisading.[13]

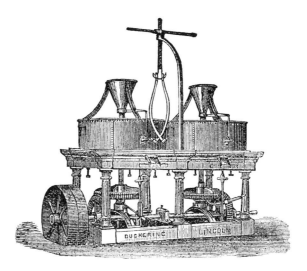

Fig 57
Double grinding corn mill made by Duckerings, 1903.

This year also saw Charles elected as Sheriff of the City of Lincoln, 'which he filled with singular distinction and ability.'[14] The following year he became a father for the fourth time when son Richard was

Fig 58
Part of Richard Duckering's Waterside North Works in 1961 after the site was acquired by YMCA.

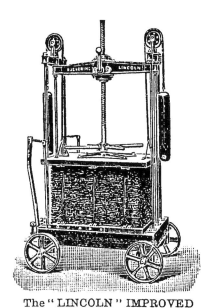

Fig 59
Straw trusser made by Duckerings, 1903.

home-made bread baked therein and exhibited by me at the Dairy Show, held at the Agricultural Hall, Islington, London N in October last. I may add I do not now use my brick oven at all, the range as affixed now answering all my purposes. Signed, Wm C Dixon, Thorpe on the Hill, Lincoln, dated December 15th 1893.'

More publicity from 1897 has a line drawing showing two maids busy on washday with a 'Beard's Challenge Laundry Stove' in the foreground *(Fig 60)* with the declaration: 'Defies Competition, None genuine unless marked 'The Challenge', saves fuel and labour, thousands sold - prices from 35 shillings, made in 5 sizes, warehouse and stores at 27 Silchester Road, Notting Hill W. Manufactory - Waterside Works, Lincoln. A month's free trial if desired.' This would appear to be Duckerings' outlet in the capital.[17] In 1899 Duckerings could be contacted by national telephone No.63. 1901 saw them exhibiting their wares on market stalls every Friday on Lincoln's Cornhill and every Wednesday and Friday at the Cattle Market on

born, a third daughter having arrived in 1873. The 1881 census tells us that as a two year old 'nursed child' Richard was boarding with a sheep-dipper named Henry Hibdon at Heighington.

1879 was a significant year. A new foundry was built on Rosemary Lane and the firm won the prestigious contract for the structural ironwork at Lincoln's Corn Exchange in the city centre. In 1881 they were the sole manufacturer of the 'Premier Range' (kitchen range), with the boast that it 'can be used as an open or closed fire at the cost of 4d per day.'[15]

The Institute of Mechanical Engineers accepted Charles Duckering as a member in 1885 and in their journal of the same year said of his foundry: 'These works … have continued to grow steadily and now embrace the business of a general engineer and millwright. The number of hands employed exceeds 80 and the works cover a tolerable area, with ample space remaining for further development'.[16] The firm's address became 'The Waterside Works' and was extended to general engineering; sheds were built off Sparrow Lane later in the year *(Fig 58)*.

An unsolicited testimonial appears in an advertisement from 1894 stating: 'The range you supplied me with for my farmer's kitchen in January of this year gives me great satisfaction and I have further pleasure in stating I have received the silver medal (first prize) from the British Dairy Farmers' Association for

Fig 60
Laundry stove advertisement, 1897.

Fig 61
Duckerings first used this Trade Mark in 1903.

Monks Road. By the 1900s they were making corn grinding machines in substantial numbers. 1903 also saw the first appearance of their trade mark - a duck as the emblem on a signet ring *(Fig 61)*.

The 1907 Royal Show at Lincoln

In June 1907 King Edward VII visited the Royal Agricultural Show held on Lincoln's West Common. Charles Duckering's company was featured as one of six of Lincoln's leading manufacturers in the agricultural trade included in the official Royal Show Guide, produced by the *Lincolnshire Echo*. A short history of the firm was provided together with a testimonial to Mr Duckering and his photograph. It also gave a description of the company's pattern department which began:

> 'A feature of the works is the pattern department. A great room on the north side of the works is filled to the doors, but this is not nearly sufficient space, and the patterns overflow into other shops near the main gate and on the ground floor. There have been absorbed into the extended premises a number of former cottages to the south-east and these again are stored to the roof with patterns, while others abound in buildings running the length of the foundry's eastern boundary. It should not be supposed however that these patterns are simply shelved as they are finished, and forgotten, to increase item by item a cumbersome and useless store. The real fact is that these patterns are valuable and are annually overhauled, those not called into use or likely to be needed being promptly destroyed. Thus the long range of patterns to be seen on the premises speaks more clearly and emphatically than could any other testimony of the position held by the firm and the system by which its business is carried forward. Since Mr Duckering took over the reins the firm has vastly expanded its area of influence, increased its number of employees, and kept thoroughly up to date in the style of its output.'[18]

The Ruddock's edition of the Show guide has a good advertisement with a clear and prominent trademark. Their introduction to Mr Charles Duckering's Waterside Works contains an interesting description of the factory as a whole:

> 'The output of these works falls, broadly speaking, into four sections. Firstly there is the foundry, the castings of which right from the beginning have earned and maintained a very high and enviable reputation. Secondly, there is the engineering department, where the famous corn mills and other agricultural machines emanate. Thirdly, the builders' department, a line which has ever been specially catered for at Waterside Works. Here practically every kind of ironwork used by builders from damp course to the roof is supplied. Under this heading particular notice must be made of the firm's cooking ranges, stoves, mantle pieces etc., which have long been a special feature of the business, and, we believe, with regard to the immediate neighbourhood at any rate, we are correct in saying that those houses are in a decided minority in which there is not a Duckering Kitchener. For the better exhibition of its ranges, stoves, mantle pieces, etc, the firm has just opened new and extensive showrooms on Monks Road, within a stone's throw of the works. And fourthly, reference remains to be made to the sanitary ironwork turned out at these works, this department being well laid out to deal with the requirements of Corporations, Rural District Councils, Sewerage Contractors etc. In fact the largest and most varied selection in this department can be found of any in the county, and compare favourably with any other of its class in this country.'[19]

Fig 62
From an advertisement in Ruddock's Lincoln Directory, 1901.

It seems that by 1907 Duckering's business was able to supply a broad and continuous range of items for many purposes and trades.

Around this time the firm was also supplying the castings to Lincoln City Council for use in the construction of seven streets of new houses on the northern side of Monks Road, east of the City centre. Many examples of different items *(Fig 64)* can still be seen in these streets today.

In 1909 the company reached the impressive milestone of manufacturing and selling 30,000 of its kitchen ranges, many fitted in the new houses that were being built in the City at the time. Four years later they had produced another 5,000, an average of about twenty-five every week, in addition to all their other products.[20] The Excelsior Range, one of the most popular models, was almost 1 m (36") wide with skirtings, fire, boiler, ovens and sham side. It weighed 150 kg (3 cwt).

The Duckering foundry passed into its third generation when grandson Richard joined the firm as manager after his father's retirement in 1912. He was paid a salary of '£500 per annum plus £10 per cent of annual profit',[21] later inheriting the business when his father died. Charles had had more than 50 years' experience in iron founding and it was said that 'there was at least one machine still in use at the Waterside Works that he had helped mould the cylinder for'.[22] His secret of success was 'to individualise his customers and to make a special and continuous study of the requirements of each, and, because of this constant attention and his untiring labours, the business grew. It vastly expanded its area of influence and had increased its number of employees to over one hundred men and boys [by 1907] under his guidance'.[23]

Charles Duckering had spent his life as a member of the Free Methodist Church and had held several posts in the movement. He served as Sunday School Superintendent for 45 years and filled the office of President of the Sunday School Union four times between 1892 and 1897. He was an overseer of the Parish of St Swithin for twelve years, the parish in which he had lived for all but nine months of his life.[24] In fact he lived no further than 800 metres from his Waterside Works in all that time. He was widowed in 1915 and moved to South Scarle, Nottinghamshire, in the spring of 1916, where he died of acute bronchitis on 17 December 1916, aged 75.

Fig 63
Duckering's shop, 29 Monks Road, from a 1913 Directory advertisement (In 2006 St Barnabas Hospice charity shop).

Charles had received at least two awards for the company's products. In 1891 he was awarded a gold medal at the Jamaica Exhibition for his celebrated 'Lincoln' Corn Grinding Mill[25] and a 'Certificate of Merit' at the Lincoln Health Exhibition of 1899 made by The Sanitary Inspectors Association for 'General Excellence of Exhibits'.[26]

Duckerings' contribution to the 1914-18 War effort was to produce small castings for William Foster's newly invented Tank. Business flourished during this time and the workforce rose to several hundred, but the conflict created personal problems for Richard Duckering's wife, Marie. She had been born in Germany and had to go to some trouble to disguise her nationality.'[27] Following the end of the hostilities, the industrial slump of the 1920s hit Duckerings hard and they had to seek other work. Employee numbers fell drastically to about thirty contracted people. This among other factors led Richard, in early 1920, to turn the foundry into a limited liability company.

Richard's son, Dick, suggested in a letter written in 1983 that his Grandfather, Charles, was not a good businessman. 'He acquired many small houses that soon degenerated into slums, he gave away every Christmas large quantities of poultry, pheasants, etc, and ran up a large overdraft. When Charles died Richard inherited the business and the debts.'[28]

In October 1920 the Lincoln Branch of The Institute of Mechanical Engineers published their journal in which they included a brief report of the history of the firm, stating that: 'From the beginning the foundry has formed the chief portion of the business and is at present being extended to meet the constantly increasing demand for high class iron castings, for which the firm have always had a good reputation'.[29]

Around 1925 fire grates were the latest items from the Duckering foundry to appear in their showroom. They were made in various sizes from 'wee cottage' to 'large farmhouse fires'. These fires had a side boiler with tray door, which was filled using a 2-gallon spout can - a very short can with a 5cm diameter spout. It had a grate in the middle with a cast-iron warming rack above the grill. These were made until at least 1947.

Company Changes in the Twentieth Century

In 1926 Richard Duckering Ltd collapsed when 'orders were few and far between'.[30] Possibly Richard was unable to get any more credit because of the debts left by his father. But he now took up a partnership with a Mr Worley, travelling to major ports around the world introducing a system of simplifying code books. He managed to pay off the foundry's debts eventually, but had to sell the family home, 'Nelthorpe', on Wragby Road, (and possibly also their shop on Monks Road). By this time all the family's wealth had evaporated[31] and the family moved to Cambridge, where Richard died in 1964. None of his children married and there are no direct descendents of this branch of the Duckering family.

However, the business was continued by the works manager, P G H 'Percy' Freeman, who took over the foundry and retained the Duckering name. Much of the work being done at this time was concerned with electrical generators for a company (thought to be Ferranti). Then in 1930 Clayton and Shuttleworth ceased trading and their steam engine drawings, patterns and templates were taken to the Waterside Works, where Duckerings undertook to supply the spares for Clayton's railcars and shunting locomotives.[32] Due to the collapse of Clayton's Wagons, no further development of their steam locomotives or railcars was possible, so unfortunately no significant extra work came to Duckerings from this venture.

Percy Freeman has been described by a former employee as 'a good, unassuming, quietly efficient businessman whose watchword seems to have been "Quality". If one casting out of a hundred was not up to scratch he would want to know why. He would regularly travel up to the north-west in the company's Austin 16 to Manchester Dynamo and Crypto, Duckerings' main client at the time. Mr Freeman would wear a shabby suit and travel up to Manchester, stay over for two or three days, wine and dine their management and buyers and return with armfuls of blueprints for work. We used to think he was having a bit of a holiday when he went up there'.[33] A lorry was later purchased to deliver these goods.

Drainpipes, guttering, drain covers, inspection covers, gas-light columns and iron bench seats were also being made at this time, and 'the spare storage sheds were still full of the old patterns'.[34] An air raid shelter that had been erected in 1943 was converted into a drying and washing room in 1947. Land was also purchased in this year which was used to build a new pattern shop twelve months later, at a cost of £2,514. Two extensions were made to the moulding shop in 1954 and 1955 at costs of £2,580 and £1,600 respectively.

Percy Freeman died in about 1952 and his wife took over the running of the foundry for a while. Later, the company employed a manager, who moved to Lincoln from Brighton in late 1953. He introduced a new system of producing castings by gas moulding in about 1955. 'This turned out to be a total disaster. The manager was 100% committed to the new method and we had to put up with it. The method involved altering all the original pattern boxes, which was irreversible once done. Holes were drilled into the sides of the boxes for the gas pipes; this froze the sand, but the heat of the molten metal "thawed" it. (The moulds were not baked in the ovens or coated with a "grey slime" when gas was used). And so when the finished product was removed, the sand and metal casting mixed, which the slime would have prevented. Instead of producing one bad casting out of every twenty, we were getting only one good one!'[35]

This manager also took on the regular trip to Manchester. 'He would put on his best suit and go up there and tell their management they wanted this and that and how to do things, and would come back with next to nothing. We thought differently about Mr. Freeman's trips after that'.[36]

By 1957 there were only about seven members of staff employed working a 44-hour week, starting at 7.30am until 5.30pm. 'The working conditions were depressing, partly due to the rumours of the imminent closure of the factory'.[37] Staff also left through boredom due to lack of work. The only work coming in towards the end of the 1950s 'were a few sympathy orders from

Sewell Road

Becor House

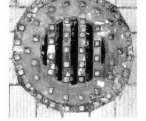
Clasketgate

Russell Street

Horton Street

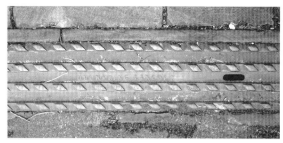
Minster Yard

Little Bargate Street

Wragby Road

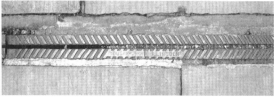
Greestone Place

Coulson Road

Lincoln Castle

Fig 64
A selection of street furniture manufactured by the Duckering foundry and still to be seen in the city of Lincoln today.

Manchester Dynamo and Crypto and a few repeat orders for drain covers. The foundry closed due to lack of orders, lack of good management and old fashioned machinery'.[38]

The closure was announced in the *Lincolnshire Echo* on 22 June 1962:

LINCOLN FIRM IS TO BE WOUND UP

The engineering and ironfounding firm of Richard Duckering Ltd, Waterside, Lincoln, is going into voluntary liquidation. An extraordinary general meeting of the company has resolved that the company be wound up. A declaration of insolvency has been filed estimating that all creditors will be paid in full in due course. Appointed joint liquidators are Percy Cauldwell of Queen Street, Sheffield, and Philip Gordon Stone of Newland Chambers, Beaumont Fee, Lincoln. The company will shortly cease trading.

The firm was established by the late Mr Richard Duckering in 1845 on a site on the opposite side of the river to the one now in use. The move to the present site took place in the late 1850s.[39]

Notes

[1] *In the Old Armchair,* The Lincoln Gazette, (Lincoln, c1920), pp.108-111; 1851 census.

[2] *Ibid.,* p.111

[3] Mills, D R and Wheeler, R C, *Historic Town Plans of Lincoln 1610-1920,* The Lincoln Record Society (Lincoln, 2004), pp. 65, 81, 96.

[4] *Stamford Mercury,* 3 September 1847, p.3, col.4.

[5] *In the Old Armchair,* p.110.

[6] *Ibid.,* p. 110.

[7] *Ruddock's Trade Directory of Lincoln,* J W Ruddock, (Lincoln, 1848).

[8] *In the Old Armchair,* p.110.

[9] *Ibid.,* p.110

[10] Indenture at Lincolnshire Archives (Hill 28/1/5/4)

[11] *Green's Royal Lincoln Show Guide,* Lincoln Gazette, (Lincoln, 1907), pp.26, 27.

[12] Morris & Co

[13] *Directory of the City of Lincoln,* Charles Akrill (Lincoln, 1877), p.178.

[14] *In the Old Armchair*

[15] *Ruddock's Trade Directory* (1858).

[16] *Journal of the Institute of Mechanical Engineers,* (London, 1885), pp.305, 444.

[17] *Directory of the City of Lincoln,* J W Ruddock (Lincoln, 1897), p.235.

[18] *Green's Royal Lincoln Show Guide,* pp. 26,27.

[19] *Ruddock's Royal Show Guide,* J W Ruddock & Harrison (Lincoln, 1907), p.46.

[20] *Trade Directory of the City of Lincoln,* J W Ruddock (Lincoln, 1909), p.313.

[21] Will of Charles Duckering, Lincolnshire Archives (Wills 1918.598).

[22] *Green's Royal Lincoln Show Guide,* pp.26, 27.

[23] *Ibid.*

[24] *Ibid.*

[25] *Stamford Mercury,* 12 June 1891, p.5, col.1.

[26] *Trade Directory of the City of Lincoln,* J W Ruddock (Lincoln, 1899), p.275

[27] Letter from Richard Duckering, 23 June 1983.

[28] *Ibid.*

[29] *Journal of the Institute of Mechanical Engineers,* (London, 1920), p.790.

[30] Letter from Richard Duckering, 23 June 1983.

[31] *Ibid.*

[32] *Clayton Steam Wagons,* J Ruddock & R E Pearson (Lincoln, 1989), pp.37, 38

[33] Alex Wilcockson, oral information.

[34] *Ibid.*

[35] *Ibid.*

[36] *Ibid.*

[37] *Ibid.*

[38] *Ibid.*

[39] *Lincolnshire Echo,* 22 June 1962, p.1

CHAPTER 5

J B EDLINGTON OF GAINSBOROUGH
AGRICULTURAL MACHINE MAKERS

Susan Edlington, Tony Wall and Terry Maidens

The Founding of the Business

Edlingtons began in 1865 as a partnership between John Butler Edlington, aged twenty-six, and his elder brother Thomas, who was six years his senior. They were born at Bottesford, near Scunthorpe, the sons of John Edlington, a farmer, and his wife Mary. There were also three other brothers and a sister in the family. John and Thomas had first moved to Gainsborough to work for the engineering business run first by William Marshall and then, after his death in 1861, by his son James. At this time, around 1860, there were only about a score of men working for the Marshalls, but the business, under the name of William Marshall Sons & Co, eventually grew to become an international concern employing 14,000 people by 1914.

Trading as J B & T Edlington, the new business was described as engineers, millwrights and iron and brass founders. It seems clear that there was no ill feeling when the brothers left Marshalls to start their own enterprise because, in April 1880, James Marshall inscribed a copy of *The Moulder's and Founder's Pocket Guide* by Fred Overman 'Presented to Thomas Edlington in recognition of his attention to explanations respecting the steam engine and his intelligence in responding to queries relating thereto'. The brothers tended to concentrate on the lighter types of agricultural machinery such as reapers and mowers which complemented Marshall's range of products, rather than competing with them.

This is reflected in the only two patents granted to the two brothers: No.3433 (19 December 1871) 'for the invention of improvements in smutting and cleaning grain in threshing machines'; and No.3233 (25 August 1877) ' for the invention of improvements to horse rakes'.

The premises that the brothers originally rented at 24/26 Lea Road, Gainsborough had probably been previously occupied by William Cross Furley and was known as the Phoenix Works. Furley, shipbuilder and boiler and engine builder, was recorded in the *London Gazette* on 7 February 1862 as having assigned all his property to John Firth of Sheffield, steel manufacturer, Thomas Adwick Farmer of Gainsborough, ironfounder and ironmonger, and Samuel Wilkinson of Gainsborough, bank manager. This probably explains the origin of the name 'Phoenix' used by Edlingtons in future years.

Fig 65
John Butler Edlington (1839-1921) joint-founder of the company.

In 1873 the larger site at Lea Road adjacent to the Great Central Railway line was purchased from Joseph Laughton, wharfinger, and Charles Crosby Patrick, grocer. A brick tower mill, dating from 1791 and known as Cont's Mill, occupied part of the site. It was converted by the Edlington brothers to steam power. The mill was advertised in the *Gainsborough News* of 4 November 1882 for letting or for sale as 'a first class steam flour mill containing eight pairs of French stones fitted with blast and exhaust, two silk reels (Hopkinson's Patent), bran cleaner, corn screen, Child's aspirator, wheat damper and driven by two condensing steam engines and two double-flued boilers'. A further advertisement on 16 February 1884 sought 'a good Millwright accustomed to flour mill work', and then on 23 August 1884 another advertisement offered the mill for sale or to let once more.

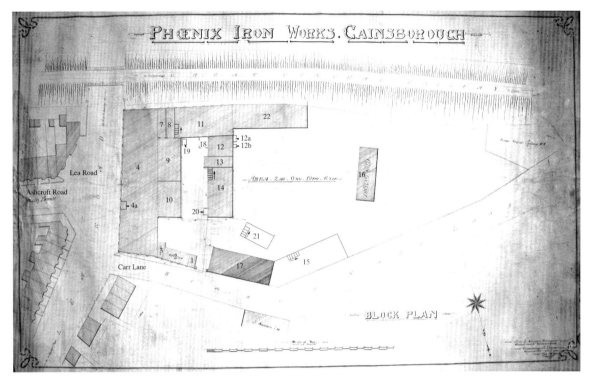

Fig 66
Block plan of the Edlingtons' works on Lea Road site in 1899.

1: Time and weighbridge office; 2: Offices; 3: Records and archive store; 4: Main workshop; 4a: Office; 7: Steam engine; 8: Boiler;
9: Steel store; 10: Parts store.11: Moulding shop; 12: Brass foundry; 12a and 12b: Coal and wood stores; 13: Coke store;
14: Fettling shop; 15: Showroom; 16: Wood drying store; 17: Paint shop; 18: Cupola(s); 19: Women's lavatory; 20: Men's lavatory;
21: Stable with hay loft above; 22: Store (timber built).

Fig 67
J B Edlington & Co Ltd letterheading with illustration of the Phoenix Works.

Fig 68
Paint shop with binder and turnip choppers visible. Glazing in the north facing roof maximised natural light.

Fig 69
Moulding shop showing moulding boxes and sand. Molten iron came from cupolas beyond the facing wall to the left of the large doors.

The first workshop was built in 1875, followed in 1883 by a machine shop, and in 1885 by a foundry with chimney, boiler and engine house. In 1886 a large two-bay workshop measuring 193 ft (58 m) by 53 ft (16 m) was erected together with offices and an entrance on Carr Lane and a large wooden shed for storing implements *(Fig 66)*.

The *London Gazette* confirmed that the Edlington partnership was dissolved on 16 September 1893. Thomas, by then over sixty years old, sold his shares to his brother and went to farm at Messingham. John, continuing to run the business as a sole trader, extended and improved the site at Lea Road by purchasing additional land.

The paint shop *(Fig 68)* was built in 1896 along with the entrance gates and time office. A moulding shop *(Fig 69)*, coke store, fettling shop and pattern shop were the next structures to be built in 1899. Following the purchase of adjacent land from Sir Hickman Bacon in 1911/12, a showroom *(Fig 70)* and loading ramp were erected.

The business expanded further in 1896 when John purchased the Victoria Foundry in Brigg from John William Spight. (Isaac Spight, John's father, a much respected iron founder and machine maker, had worked from the Victoria Foundry for almost 50 years until that date). It is thought that he had intended to build a saw mill on the site, but this did not come about. By 1904 the total workforce had grown to some eighty men.

The Formation of the Company

It was in 1903 that a proposal for the flotation of a new company, J B Edlington & Co, was made. W H Hanson JP was to be chairman and J B Edlington the managing director. The rest of the board was listed by the *Gainsborough News* as C F Hookham of Brigg, J H Skevington of Brigg, J Spilman JP of Brigg, W S Swift of Walkeringham and R Wheeldon of Eaton near Retford. W D Ogle of Gainsborough was company secretary. The prospectus valued the company at £20,000.

The limited liability company was eventually registered in November 1905, with J B Edlington as governing director, T and T E Edlington and W S Wilkinson as directors. Wilkinson had trained at Marshalls before going to Mysore in India, where he became manager of Mathesons' works and sawmill. Upon his return to England with his family, he joined J B Edlington & Co as a director and as company secretary. He was later to become managing director, a post that he held until his retirement. At the end of the first year of the company Thomas Edlington took over the reins from his uncle John, who had no male heir. Thomas was the son of John's brother William.

Annual Reports dating from the early years of the limited company give us an insight into the trading conditions faced by a company that was closely involved with agriculture during the first quarter of the twentieth century.

Fig 70
Upper floor showroom in the 1911/12 building, which replaced a structure destroyed by fire.

In 1907 a great increase in the cost of raw materials coincided with increased competition from foreign manufacturers. Three new moulding machines were installed in an attempt to reduce labour costs. Nevertheless, a significant increase in turnover kept the workforce very busy.

A small surplus was made in each of the subsequent two years, after taking into account further investments that had been made in additional machinery. The report for 1909 also mentions increased representation at many agricultural shows throughout the country as a means of advertising the products and also of developing relationships with customers and potential customers. Attendance at such shows remained a very important part of the company's activities.

The year 1911 was a bad one for farm machinery manufacturers. Climatic factors reduced the demand for food preparation machinery, especially turnip cutters. On 29 July a disastrous fire destroyed the store that was adjacent to the Great Central Railway line. The fire happened during the Workmen's Trip Week, when the factory was closed, so it is not unreasonable to suspect that sparks from a passing train might have been the cause. On a brighter note, patterns had been made for a new reaper, which was to be introduced for the 1912 season.

The new showroom and store, commenced during 1911, was finished during 1912. Potential profits for the year were largely offset by an increase in the cost of raw materials and also by cheap imports from the USA. 1913 was another poor year for farming and by 1914 the Great War had led to increased costs and reduced sales. As a result it became necessary for short-time working to be introduced.

Business improved in 1915, as efforts to maximise food production during the war had led to an increase in demand for machinery. The loss of many skilled hands to the army was a serious problem and overtime was worked throughout the year. Demand remained good in the following year and work continued on both agricultural machinery and munitions work, though no information is available about the nature of the latter.

In 1919, the first year after the Great War, there was a steady increase in trade, but raw material costs remained high and British firms found it difficult to compete in foreign markets.

A new Cornish boiler was obtained from Marshalls to provide power for the factory. A fire in the roof of the pattern store, which was above the boiler house and engine room, was promptly attended by Newsums' fire brigade, who assisted the local authority fire brigade in containing the fire and bringing it under control. (Newsums had a large timber warehouse nearby alongside the Trent. This was the first call-out for the town fire brigade for eighteen months). Extensive damage was done to the pattern store and to the boiler/engine house. Sparks from a passing train were again thought to be the cause. It had been announced in the 1919 report that the Brigg premises had been disposed of, but the 1920 report confirmed that completion of the sale, to the lessees, was not due to take place until 31 December 1922. The prevailing depression had affected trade, and short-time working was re-introduced during 1921.

Fig 71
Stand at the Lincolnshire County Agricultural Show at Scawby Park (near Brigg) in 1953.

Fig 72
The workforce just before WWI. Thomas E Edlington is seventh from the left on the front row.
The bell on the roof was rung at the beginning and end of shifts.

J B Edlington, the founder of the company, died on 9 January 1921 at the age of 81. He had remained active to the end of his life, having designed a successful new potato sorter in his seventy-second year. 1922 was the worst year for trade since formation of the company, due to the precarious state of agriculture in the United Kingdom. The lessees of the Brigg premises had given up their option to purchase the building, so the company resumed the business there, operating it as a retail outlet. The following year a profit was made and stocks which had been built up over the previous three years were sold. The increase in sales was, in part, due to the success of the Brigg branch of the business.

Business continued to improve over the next two years but then, in 1926, trade was affected by the Coal Strike. However, the 'Gainsbro' High Speed Grass Mower won a silver medal at the RASE Show at Reading and this encouraged more buyers.

The Next Generation Joins the Board

Another generation of the Edlington family joined the board in 1926 in the persons of Thomas and Jack, both sons of Thomas E Edlington. The sales of grass mowers increased as a result of the silver medal, but sales at Brigg were not so good, because 1927 was a bad season for farmers in Lincolnshire.

Sales of the 'Gainsbro' Mower were up by 50% in 1930 and the new beet elevator and cleaner won a Gold Medal at the Sprowston Beet Trials in Norfolk. However, the price of potatoes had dropped dramatically and so sales of potato sorters were down by 35% and of turnip cutters by 50%.

Bad debts among customers were higher than normal and these trends continued through the next three or four years. Very many companies of a similar size to Edlingtons succumbed to the effects of the depression in the late 1920s and early 1930s, but

Thomas E Edlington's enthusiasm for innovation knew no bounds and the company survived as a result of his inspiration and leadership.

Thomas E Edlington died in 1934 and his third son Alfred then joined the board. Alfred ran the works, Jack was responsible for finance and administration and Thomas designed and sold the machines. A new potato sorter was introduced and the potato digger sold well. The following year the potato digger won a silver medal at the Lancashire Show and a new self-locking steerage hoe generated much interest. This was also the period when the firm developed tractor drawn machines for sowing and harvesting potato and sugar beet.

The improvement in the company's fortunes continued through 1937 and 1938. In the latter year, some of the patterns, jigs and templates from the late business of John Cooke & Sons Ltd of Lincoln were acquired. When the Second World War started in 1939 some large orders were received from the Ministry of Agriculture. As in the Great War, there was increased demand for agricultural machinery, but the number of skilled men being called up for military service caused great difficulties in the maintenance of production. Even though women were employed, fuel, gas, electricity and water all increased in cost by 50% and cut into profit margins.

The lack of available skilled men continued to be a problem in the post-war years. In 1957 Mr Quickfall, the manager at Brigg, died and Brian Edlington

Fig 73
Brian Edlington (1928-1997) in 1992.
Chairman and MD following the death of his father, he was responsible for transforming the company in the 1960s.

Fig 74
One of the many medals won by Edlingtons at agricultural shows and exhibitions over the years.

(Fig 73), the son of Jack, took over his role. Although Brian was successful in increasing sales at Brigg, losses in other directions often offset this gain.

By the time that his father died and Brian had taken over the running of the whole business, things were in a fairly poor state. Brian realised that the company needed new modern designs and the capital to build the machines. The traditional materials then being used had to be replaced, and all-steel products offered to the customer. That Brian was successful in turning the business around is evidenced by the fact that Edlingtons are still in business, unlike so many of their former competitors. Brian exhibited a similar flair for innovation that had been seen in his grandfather Thomas, and a variety of new machines, including some very large beet handling lines, were designed and built. He also designed a machine that sifted and mixed soil for making and repairing the greens on golf courses; one was purchased by the Woodhall Spa club, but unfortunately further orders were not forthcoming.

Paul Edlington, Brian's son, is now the managing director of the company. Paul's mother Susan is a director and the company secretary.

The Products

Advertisements in the *Gainsborough News* during 1875 indicated that the company were 'Manufacturers of Patent and Improved Reaping and Mowing Machines, Hay Making Machines, Horse Rakes and Improved Horse Gear, Chaff Cutters and Grinding Mills'. In addition, work was done for the local waterworks. By January 1876 the range of machinery had expanded to included turnip cutters, root pulpers, Cambridge rolls, harrows, steerage horse hoes and corn and small seed drills.

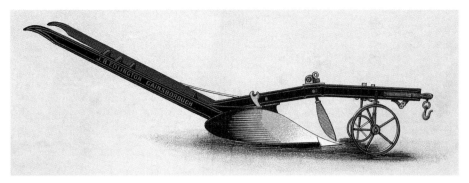

Fig 75
Plough c1880. Early Edlington advertisements offered a dozen different types of plough and marketed those of other makers.

The fact that the Edlington brothers intended to make the manufacture of the lighter type of agricultural machinery the basis of their business did not stop them from undertaking other work. This is indicated by a report of 11 November 1876 which stated that J B and T Edlington had just completed at their Lea Road works the manufacture of probably the largest engine ever made in Gainsborough. It was a horizontal, expansive, condensing steam engine of nominal 50 horsepower and fitted with all the latest improvements. The destination of the highly finished product was Glasgow. The purchasers, Messrs Pearson Bros, had contemplated the erection of an additional oil cake mill in Gainsborough, but it was later decided that Glasgow would be a better place for the undertaking; hence the engine was to be sent to Mr J Pearson's Rockvilla Oil Mills in Scotland. In July 1882 Edlingtons exhibited 5 hp and 2½ hp vertical engines as well as various other types of agricultural machinery.

No details have been traced to indicate volume of sales for individual products in the nineteenth century, although catalogues for the annual County Show give details of the wide range of machines they made. In 1902 sales were recorded as follows: 183 mowers; 121 horse rakes; 65 self-delivery reapers; 4 self-binders, 417 ploughs; 262 turnip cutters; 42 pulpers and graters; 25 chaff cutters; 52 cake mills; 3 grinding mills; 77 spring-tine cultivators; 34 harrows; 52 corn and turnip drills; 36 Cambridge rolls; 14 weighers; 15 hay collectors; 31 horse hoes; 10 horse and pony gears. In addition, orders were on hand for 69 grass mowers; 3 reapers; 12 self-binders; 1 haymaker and 37 horse rakes.

A major development in 1911 was the 'Phoenix' cylindrical potato sorter, which was to retail at £7.15s. A contemporary report of the trials read: 'Messrs Edlington's machine finished with the help of five men and no manager in 45 minutes, although the breakage of a linking chain delayed them a short time. Curiously enough an assistant was at once seen to be leaving the pink ones amongst the seed, but, on colour blindness being found to be the cause, an exchange of attendants soon put matters right. Their rotary screen was supposed to take out the seed and chits. At the pace attempted a good many slipped past into the waste. An especially good point in this machine was the room and facilities given for sorting diseased from the seed as they fell through the riddle.'

A power-driven version of the 'Phoenix' potato sorter was introduced in 1909 at the price of £70. The Royal Show at Derby in 1921 saw the launch of a power-driven turnip cutter - Gardner Pattern - with 45 knives, at a price of £85.

A revised price list issued on the 14 May 1919 gave details of the 'Phoenix' and 'XL' reapers *(Fig 76)*,

NEW PATENT XL SELF-RAKING REAPER.

Fig 76
Mowers and reapers remained among the company's best selling products right up to the 1950s.

the latter having two extra options in the form of a 5 ft 6 in cut instead of the standard of 5 ft and a tubular shaft. Three different mowers were offered: No.8 Phoenix - right-hand cut, No.9 Phoenix - left-hand cut and the 'U' 2-speed mowers with either left- or right-hand cut. The optional extras were a 5 ft cutting bar instead of the standard 4 ft 6 in.; reaping attachments for any of the above; pea-lifters for cutting peas or laid corn, and a wrought-iron knife stand. Twenty-five different types of plough, plus four optional extras, were listed. There were three manual-delivery horse rakes, each of different widths and with different numbers of teeth. Under the heading of food preparation machinery there were four single-action and one double-action turnip cutters, four graters, four chaff cutters, three cake mills and two potato sorters. Turnip and mangold drills of either 'Yorkshire' (with wheels) or 'Scotch' (without wheels) types were listed, with optional side hoes for either type. There were land presses with either two or three wheels. Land presses with corn drill attached came in two, three or four wheeled options. Fifteen pulpers and graters; three chaff cutters. The final item on the list was a driver's seat which, as an extra, cost £1.

Analysis of the accounts for 1919 indicates the sale of the following numbers of machines: 8 reapers, 187 mowers, 558 ploughs, 46 horse rakes, 230 turnip cutters, 15 pulpers and graters, 3 chaff cutters, 9 cake mills, 323 potato sorters, 27 corn and turnip drills, and 77 cultivators. Grass mowers, potato sorters and ploughs thus continued to represent the core of the business.

In April 1924 *The Gainsborough News* reported on 'an ingenious invention from Messrs Edlington.' For the previous 18 months they had been manufacturing the Watson Patent Incinerator. In pre-World War I days, Watson, the inventor, was the chief designer for a leading manufacturer of incinerators. Five sizes were offered, the largest of which was suitable for the destruction of rubbish for a town of 5,000 inhabitants. Local authorities all over the United Kingdom were interested and examples were also supplied to a number of famous sporting venues, including Wembley, Wimbledon, Lord's, The Oval and Tottenham Hotspur's White Hart Lane.

By 1927 the best selling product was the 'Gainsbro' High Speed Mower and many orders were received in an otherwise sluggish market, although the 'Phoenix' Potato Sorter *(Fig 77)* also sold well. In 1930 it was reported that, taking advantage of an increase in the sugar beet industry, Edlingtons had introduced a machine to clean and load sugar beet. In 1931 this machine won silver medals at the Leicester and Peterborough Shows. A gold medal had also been awarded by the Norfolk Agricultural Society following trials during which the machine topped, cleaned and loaded into carts more than two acres of sugar beet per seven-hour working day at a cost of 13 shillings, or less than 6 shillings per acre, a saving of 2 shillings per acre in this particular work. The incinerators continued to sell well to municipalities and hospitals as well as private companies, and the company reported the best trading year since the 1914-18 War.

A vertical link-action two-speed potato digger was added to the range in 1933. The new machine won a gold medal at the Holland show and silver medals at four or more county shows. Demand for implements increased during 1934 and products designed for the beet grower were especially well received. By 1936 orders for the new potato digger were exceeding expectations and development of the self-lock steerage horse hoe was recognised when the implement was held to be the most improved piece of equipment at the Devon Show. Sales of mowers to overseas markets, New Zealand in particular, increased and as a result of all these factors full-time working was maintained.

The coming of the tractor revolutionised the class of equipment used on farms and Edlingtons endeavoured to keep up with developments by designing special implements for such work. An example was the tractor roll, made in three sections and capable of rolling any width up to 24 ft, which enabled the farmer to roll upwards of 100 acres per day. The potato grower was also catered for with the introduction of the No.7 Potato Sorter in 1938. In that same year the 'Cooke' ridger, coverer, cultivator and steerage hoe were re-introduced as parts of the Edlington range. Based on the patterns and designs acquired from the Lincoln

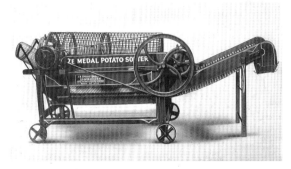

Fig 77
Phoenix potato sorter c1911. It could be driven by petrol or paraffin engine.

Fig 78
Potato sorter or grader 1966, one of the first new designs by Brian Edlington.

firm of John Cooke, which closed in 1938, some of these tried and tested implement ranges were made into the 1950s.

Throughout the Second World War the works were kept very busy with large orders from the Ministry of Agriculture in addition to the normal trade, as efforts were made to produce as much food as possible in the UK. As was the case in the Great War, a problem was encountered when numbers of skilled employees were called up for military service and it was not until 1947 that new products were launched. The 'Gainsbro' four-row drill was for sugar beet and light seeds. A six-row crop drill, developed at the request of customers for use with tractors, had also been displayed at the Royal Show at Lincoln in June.

The 'Phoenix' root washer appeared at the 1951 Royal Show and the 'Phoenix' carrot grading machine at the 1954 event at Windsor. By the latter year there was a noticeable reduction in demand for certain products as they became out of date as a result of the increasing use of tractors. The new No.10 potato sorter, introduced in 1959, was equipped with flat riddles and a feeding elevator and bagging elevator. It could be used for ware or seed potatoes. The 'Phoenix' carrot washing and grading machine, which sorted carrots into three grades, was designed to tie up with the new pre-packaging units that were being installed by customers.

Continually striving to improve the products, the firm introduced the Mk12 potato sorter in 1966 *(Fig 78)*. This sorter, or grader, was a great success and the following year a version described as the Mk3 appeared. The model number was intended to reflect its smaller size and to appease superstitious customers. The Mk14 was a grading line dating from 1969 which could handle 50 tons of potatoes. This was the year in which the first hopper/elevator was designed and built.

In 1975 two red beet (beetroot) handling lines were built to order; one was fixed and the other mobile. The second mentioned was, at the time, the largest mobile red beet handling line in Europe *(Fig 79)*. Towards the end of the decade large orders were received from the food processing industry for equipment where potatoes straight from the field could be converted into packaged frozen chips.

The 1980s saw the introduction of hydraulically operated Cambridge rolls, designed by Brian and designated HFR (hydraulic folding rolls). These had the capability of rolling widths up to 12 m and ensured that the company kept pace with customer requirements in all sectors. Further developments of the HFR into the even larger and more technically sophisticated HFRAW-HD and FB rolls have continued this trend right up to the present day.

Fig 79
70ft Mk16 Red Beet Line of 1975. It was the largest in Europe at the time.

CHAPTER 6

T & J FLETCHER OF WINTERTON

AGRICULTURAL IMPLEMENT MAKERS

Charles Parker

Beacock and Fletcher: The Early Days

When John Fletcher moved to Winterton in the early part of the nineteenth century, few people could have expected that the family business which he established would still be involved in agricultural engineering over 170 years later.

At that time Winterton was a small market town in North Lincolnshire with a population of about 1,300 and the area was heavily dependent on agriculture. John was born at Haldenby Hall near Luddington at the north end of the Isle of Axholme in 1820, the second son of John and Hannah Fletcher, who farmed land owned by the Duke of Arundel. Around 1834 he moved across the Trent to Winterton in order to start work as an apprentice to Matthew Beacock and initially he lived at his employer's premises in Town Street (now Park Street), along with two other apprentices. Beacock was primarily an implement maker, but he was capable of handling a wide range of commissions, which included the manufacture of the working parts for Winterton and Appleby church clocks. (The mechanisms were probably built by William Godfrey, a local clockmaker, who had to contract out the making of the castings and frames.)

In 1840 John went into partnership with his employer and they traded as Beacock & Fletcher. (One of their earliest machines, a winnower used for cleaning corn, was found a few years ago in a barn at Barnetby and is now part of the collection at Normanby Hall Farming Museum near Scunthorpe). In 1848 he married Maria Scholey from Eastoft Grange near Luddington and they went to live at Holly House on Winteringham Road, Winterton, where his sons, Thomas and John, were born in 1850 and 1852 respectively.

When his partner Matthew Beacock emigrated in 1852, John Fletcher had hand bills printed *(Fig 81)*.

John Fletcher: The Newport Ironworks

The outbuildings at Holly House were extended to provide bigger workshops and in the same year he took over Newport Ironworks, which was located further up Winteringham Road. By this time he employed three men and four boys for the implement business. (It is also clear that he employed a gang of men to run a travelling threshing set which worked on local farms through the autumn and winter months). The name 'Newport' implies Roman connections and this particular example was a hamlet on the road to the village of Winteringham (which is at the northern end of Ermine Street where it meets the Humber). A Roman villa was found about two fields away to the south-east of the Ironworks and in August 1868 the *Lincolnshire Chronicle* reported the discovery of a Roman pottery kiln just across the road.[2]

The building had originally belonged to John Naylor, another machine maker, and parts of it are thought to date back to the Napoleonic War period. Naylor had carried out substantial improvements to the original stone building, adding an extension using bricks from the local brickworks. He also fitted Georgian house style windows, so the general public would not recognise that it was a workshop. In later years, when several wooden beams were removed from the centre section of the workshop extension in order to accommodate threshing machines for servicing, a crown and the inscription "GR IV" were found stamped on the end

*Fig 80
Portrait of John Fletcher (1820-1895), founder of the firm.*

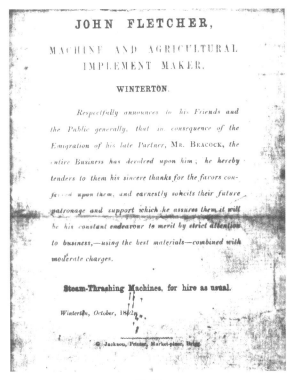

Fig 81
Hand bill distributed by John Fletcher in 1852.[1]

Newport to power the machines (lathes, drills, saws, etc) via a series of line-shafts and belts. This remained in service until 1925, when it was replaced by a 10 hp Ruston & Hornsby single-cylinder open-crank oil engine. This latter machine was used until the 1960s, by which time all of the machine tools were powered electrically.

In 1855 John Fletcher designed a horse-drawn grass mower that used the principle of a blade reciprocating over a series of fixed fingers and in 1864 he took out a patent for a renewable ledger plate for the fixed fingers' cutting edge *(Fig 82)*. (This system is still used in today's modern combine harvester). The first machine produced to his improved pattern was tried out in John Blanchard's field in Winteringham Road, Winterton. In 1870 he turned his attention to corn milling and designed and patented a 'steel plate, corn grinding mill', which was much more compact than the heavy cumbersome stone mills used in windmills and horse works up to this point. This machine was awarded a silver medal by the President of the Lincolnshire Agricultural Society at its Brigg Exhibition in 1871.

of one of the original timbers. From this it is assumed that the woodwork must have been installed in the reign of George IV (1830-37).

The development of steam power began to have a major impact on agriculture from about the middle of the century. Prior to this, most corn was harvested by hand using scythes and, after standing in stooks in the field to dry, was then threshed before being taken to local millers for grinding, although some larger farms had installed both fixed threshing and milling machines in their own barns, which were powered by horse engines. In the middle of the nineteenth century portable steam engines and threshing machines were developed, which enabled contracting teams to go around farms and thresh the corn in the autumn and winter. Grain could then be stored and taken for milling when it suited the grower.

Clayton, Shuttleworth & Co. of Lincoln were among the pioneers in the manufacture of portable engines in Lincolnshire and in 1850 Beacock and Fletcher were one of their first customers when they bought an example of the 'new improved model' for their threshing gang.[3] About ten years later, a static 7 hp Marshall (of Gainsborough) steam engine was installed in a new engine house adjacent to the workshops at

Fig 82
Plates attached to Fletchers' mowing machine and corn mill advertising patents awarded to the firm.

The mill's features were listed in a contemporary catalogue:

> These mills are entirely a new invention, and have proved themselves the best, cheapest, strongest, simplest and most durable Mills ever brought before the Public, as the Grinding Parts are made of Hardened Steel. There are only two bearings that require oil; no cog wheels, and only 1 Screw to regulate the grinding parts, which can be done at pleasure, without stopping the Mill; it will grind to softened meal for making bread or kibble as required. It also possesses great advantages over all low portable mills, as it elevates the ground corn into bags; the spout being of sufficient height, so saving a great amount of labour, the Hopper being the height of an ordinary Dressing Machine, and does not require much power to work it, as the speed required is very moderate.
> It has now been well tested on all types of corn, and has far surpassed the most sanguine expectations of the Inventor. For grinding Indian Corn or Maize it has not an equal, and can be worked by either Horse or Steam Power, and can be seen at work almost any day at the Manufactory, WINTERTON, near BRIGG, Lincolnshire.
>
> **PRICES**
>
> | No. 1 | 8 | Bshls. Per hour | ...... £12 0s. |
> | No. 2 | 10 | " | ...... £14 0s. |
> | No. 3 | 12 | " | ...... £16 0s. |
> | No. 4 | 14 | " | ...... £20 0s. |
> | No. 5 | 16 | " | ...... £23 0s. |
> | No. 6 | 18 | " | ...... £26 0s. |
>
> The above Mill can be mounted on Wheels, &c., for Travelling, for Five Pounds extra, and can be taken from place to place with one horse or pony, and put down to work in a few minutes, or can be worked on the wheels if preferred, from any Portable Engine.[4]

The other products available around this time were also listed in the illustrated catalogue; these included a 'newly invented, superior horse hoe', improved straw cutters and various types of seed and corn drill, in addition to the patented mowers and mills. He also offered combined steam threshing machines (portable and fixed) and a wide range of other agricultural equipment and products. (Surprisingly, the firm's advertisements and catalogues continued to include horse works until after the First World War, when one would have expected them to promote more modern methods of working).

Fig 83
Morris & Co Commercial Directory for Lincolnshire, 1863, advertising the wide range of Fletcher's products and agencies.

It is not known if John ever built a threshing machine to his own specification, but he did enter into correspondence with Marshalls of Gainsborough concerning his ideas for the design of straw self-feeders compared to the ones used by their machines.
It is thought that his first threshing gangs used the early type winnower machines manufactured by Beacock rather than the larger machines used later on.

John travelled around the local agricultural shows and markets to make contact with potential customers and there is evidence that he went as far afield as Grimsby, Spalding and Stamford, although his main customer base was in the area immediately south of the Humber and in the Isle of Axholme.[5]

T & J Fletcher:
The Firm Expands and Thrives

In 1877 John took his two sons, Thomas and John, into partnership and they traded under the title of T and J Fletcher for the first time. The hire and contracting side of the business was developed and they continued to expand and improve their range of products to keep up with improvements in steam power. Thomas obviously inherited his father's gift for innovation, as a 1911 show catalogue lists a portable chaff cutter and a 'batting apparatus' to fit on straw binders as his own inventions.[6] He also earned a reputation as a very hospitable salesman. Customers were welcome to inspect his products at the works and, if a decent order was placed, they would be taken over to the office at Newport House, where a barrel of good quality whisky was kept. After the paperwork was completed, proceedings would be concluded with a generous tot to cement the business. 18 June 1880 must have been quite a profitable day; hay making would have been in full swing, as Thomas recorded in the Market Book:

> Mr. Thos. Doakes, Timber Merchant, Markt. Rasen. Second hand Comb. Reaper on rail at Appleby Station. Cash £16.
>
> Wray Dixon, 4 Grass Knife eyes and 2 connecting rods. Father [i.e. John senior] brought to go back next Thursday.
>
> Mr. Vickers, Messingham Field. Wants second hand reaper. Coming over some day from 10th to 14th (July).
>
> Mr. Stamp, Caistor. Come by Harrison [carrier]. 1 reaper bar & 1 box of knives to go back next Thursday.
>
> Mr. Anderson, Miller & Farmer, Waddingham. Wants good 5 ft second hand reaper. 10 to 14£. Coming over.[7]

Milling corn was another important activity. Around this time John junior took over High Mill, immediately to the south of the works, and three steam-driven mills were installed at Newport itself, powered by the fixed Marshall engine. Local growers would look out for smoke coming out of the chimney at Newport to indicate when they could take a load of grain for milling. John senior had bought the Ironworks outright in 1872 and built some extra workshops.[8] In 1884 his son Thomas bought Newport House, a substantial house just across the road from the works, together with the surrounding land which contained the Lesser Mill and several cottages. The top and sails of this mill were badly damaged in a storm and fell on to the roof of the adjacent engine house.[9] Some time later the mill burned down and for this and various other reasons, the milling business proved problematic.

John senior had retired about 1881 and after several difficult years the two brothers ceased trading as partners in 1885, although the arrangement was not formally dissolved until 1891.[10] Thomas kept the steam mills and the implement business, while John junior moved to a house next to the Ironworks and continued to trade from High Mill until his death in 1907. The mill fell into disuse and the tower was eventually demolished in 1932, while the rest of the building was converted into a house.

John senior died in 1895, aged 75. He had developed from a young apprentice living in his employer's house to be the founder of a successful and diverse business. In addition to his passion for developing new machinery, he had served as the Parish Constable. His son Thomas became a member of Winterton Urban District Council and the local Licensing Committee. However, Thomas's public standing did not protect him from being summonsed to appear before the Magistrates for driving an unlicensed vehicle on the public road. In 1900 he was collecting a new Ransomes traction engine from Appleby station for a customer when a zealous constable noticed the lack of number plates and stopped him. Thomas was under the impression that he had cleared the delivery with the

Fig 84
Thomas Fletcher (1850-1925).

local police sergeant beforehand, but it took a considerable amount of correspondence and an apology before the matter was formally dropped! He also acquired one of the first motor cars in Winterton. Accompanied by George Waterlow, the local garage proprietor, he went to Manchester where they picked up a Model 'T' Ford and drove it back. The round trip included an overnight stop and took two days in total.

To broaden the scope of the business, several farms and other properties in the area had been acquired and these were let out to tenants.[11] Farming was relatively prosperous from 1850 to 1875, but between 1879 and 1892 rents fell by up to 60% on some estates and net income fell even more sharply, because landlords could not cut back expenses in proportion.[12] Agricultural wage rates fell and farm workers in North Lincolnshire were fortunate as they were able to leave the land to take better paid work in the iron and steel works at Scunthorpe, or on the railways. The fall in the price of wheat was a major problem for cereal growers, but, at the same time, shortage of labour meant that mechanisation was essential and this kept up demand for the company's products. A small but loyal workforce of ten to twelve men kept the Fletcher business going. The same names appear in the wages book over several decades and, comparatively speaking, their skills were well rewarded.

In 1904, while the average wage for an agricultural labourer was about 19 shillings a week, adult workers at Newport regularly earned up to one pound and 10 shillings per week.[13]

In 1894 Thomas took on the area agency for Massey-Harris implements and began to sell and service MH binders, mowers and drills. The Massey-Harris company was formed in 1891 by the combination of two Canadian companies, Massey Manufacturing Company of Toronto and A Harris, Son & Company Ltd, of Ontario. Both of these companies were industry leaders in the production of implements and this new product range would be an important source of business. MH merged with the British tractor maker Harry Ferguson in 1935 and Fletchers retained the agency until 1960, when Massey-Ferguson reorganized its franchise system.

Portable chaff cutters were another very important product which generated a considerable amount of business for over half a century. These improved food production for horses and cattle by cutting straw into short lengths to make it more easily digestible. John Fletcher's early models were single blade, free-standing hand-powered machines, but around 1870 the design was improved by the application of steam power. Initially, the new models were fitted with two wheels, so that they could be taken around farms along with the engine and threshing sets, and then they were

Fig 85
Harry Fletcher (on the left) and his team of workmen, c1920.

progressively enlarged into four-wheeled, five-bladed machines. A 'blast elevator' was added, which used a fan to blow the cut chaff through a large diameter pipe into barns or haylofts for storage *(Fig 86)*. The last one was built in 1948 and it would appear from works records that over 200 portable cutters were made in a 60-year period. During World War One the British Army hired three Fletcher chaff cutters to help feed its horses.

The company was never large enough to justify the capital investment in its own foundry, so castings were always bought from local suppliers, initially Isaac Spight at Brigg, and later, Amos & Smith Ltd of Hull. Castings for corn drills and balers were also bought from A Bucher and Co in Sweden, who was found to be one of the best suppliers. As their products were of such high quality, they needed very little cleaning up before fitting. The first orders were placed in the 1920s and the firm still trades with them today. Timber was mainly obtained from importers at Hull or local merchants, but some English oak was bought as standing woodland and felled as required. Some machines were offered in two grades, dependent on whether hard wood or soft wood was used for the frames.

T Fletcher & Son: Agencies and Repair Work

Thomas's son Harry (Henry John, 1875-1935) joined the business when he left school and, after he had served his apprenticeship, the firm began to trade as T Fletcher & Son. Their improved model Lincolnshire turnip cutter *(Fig 88)* sold in large numbers to Wolds sheep farmers because they liked its ability to cut up mangolds and turnips even when they were frozen solid, something that barrel-type cutters could not do. Turnip cutters were made until the early 1950s and, again, they were progressively improved from hand powered to petrol engine driven. The firm also had agencies for small agricultural engines built by Petter and Lister, which were bought in lots of ten for fitting to cutters and elevators.

Harry succeeded his father in 1925 and traded as H J Fletcher. The business continued to develop and improve its products and also to specialise in servicing and repairing binders and mowers. After a short period of growth just after the First World War, farming entered another slump and the company's archives contain various letters from customers who were unable to pay their bills from season to season due to

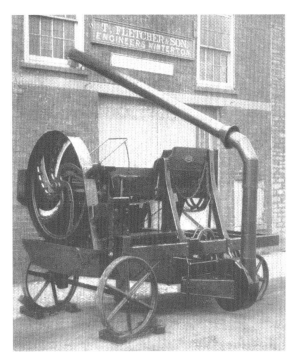

Fig 86
Chaff cutter with blast elevator, one of Fletchers' best selling products, c1920.

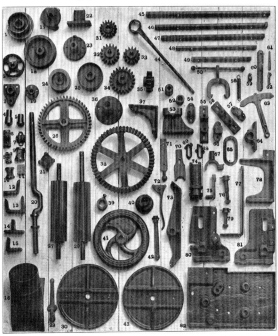

Fig 87
Replacement parts, mainly of cast iron, held in stock for the chaff cutter, c1936.

THE Lincolnshire Turnip Cutter

INVENTED AND MANUFACTURED BY

T. FLETCHER & SON,
OF THE
Newport Ironworks, WINTERTON,
Near Doncaster.

PRICE - £4:10:0.

Will cut frozen roots without any rolling about in the hopper, and cuts the last piece.
A great favourite with farmers and stock feeders.
TESTIMONIALS ON APPLICATION.

Fig 88
Advertisement for turnip cutter featuring Tom Fletcher (Jr) aged 8 (1912).

poor returns or failure of their own clientele. Fortunately, demand picked up in the late 1930s and the small size of the business, coupled with its secure asset base, allowed it to weather this second difficult period.

T & J Fletcher:
The Modern Implement Makers

With the untimely death of Harry in 1935 at the age of 60, his sons Tom (Thomas, 1904-1985) and Jack (John, 1907-92) took over the firm and the new partnership took on the title of T and J Fletcher for the second time. Tom had trained as a millwright with his father, but Jack had first trained as an auctioneer and valuer with Freddy Glasier. However, after a year's (unpaid) work he changed direction and served his apprenticeship with a local painter and decorator. Whilst the administration of the business would become his primary responsibility, his painting and sign writing skills were put to good use on the company's products and vehicles. Their other brother, Harry junior (always known as 'Soot'), first started work on local farms, but he too joined the business and worked as the blacksmith's striker until he retired in the 1970s.

In the late 1930s Tom Fletcher started another family tradition. With the threat of another war looming, he became a Special Constable and helped to establish the local Observer Corps post, which was set up to the rear of Harrison's farm quite near to the works. Jack also served on the post during the war and Tom's son John went on to become the Group Officer responsible for the posts in North Lincolnshire, serving until the ROC stood down in 1991 at the end of the Cold War.

Fig 89
Fletcher service vehicle and chaff cutter outside Newport Ironworks, 1942.

Fig 90
Aerial view of Newport Ironworks, 1970.

During the 1939-45 War spares for agricultural equipment became difficult to obtain, but the engineering skills of the millwrights and blacksmiths of Newport enabled old pre-war machines to be kept going long past the point when they would normally have been replaced. Despite the U-boat campaign, some essential parts were still obtainable from Canada. Conscription led to a shortage of manpower in the later stages of the war, although Tom and Jack were exempted because they were in reserved occupations. A German Prisoner of War camp was located at Winteringham, about two miles from the works, and for some time three PoWs, including a skilled blacksmith, were allocated to the business.

In 1949 the firm sold its first combine harvester (a self-propelled Massey-Harris MH21), ushering in a new era in farming. Improvements in tractor design in the 1950s also meant a change in the range of implements required, and cultivators and ploughs were made to suit the different hydraulic lift systems used by various makes. Rollers, Cambridge rollers *(Fig 92)*, spring tine and standard harrows (the 'Fletcharrow') replaced chaff cutters in the workshop and the company also took on the agency to sell and service Standen sugar beet harvesters.

Fig 91
Fletcher brothers: Tom (1904-85) and Jack (1907-92).

Tom's son, John T Fletcher, began his apprenticeship with the company after the war and, following National Service as an airframe fitter in the RAF, took increasing responsibility for the running of the company. John had been involved in the business almost as soon as he could lift a hammer or a spanner and, while other young boys played with Hornby trains or Meccano, his practical skills were gained from learning to dismantle and rebuild an MH binder-knotter unit at the age of six or seven. The company continued to maintain animal feed mills and one of John's first jobs was to accompany his father to overhaul and dress the stones in the mill at Robinson's Farm, Redbourne.

John took over completely when his father retired through ill health in 1965; Jack, his uncle, officially retired in 1972, stayed on to run the office.

In the 1960s larger and more efficient combine harvesters were introduced and improved handling systems were needed to manage bulk grain. The firm became an agent for Brice Baker Engineering and began to specialise in the design and installation of storage and drying equipment for farms *(Fig 93)*. By this time the older parts of the Ironworks were no longer suited to modern production techniques and a new three-bay workshop, 18 m x 7.2 m, was constructed in 1961 on the site of the old paint-shop and cart-sheds.

In 1992 John's son Richard and his wife Fran (Marie) took over the family firm, to keep up the tradition of Richard's great-great-great-grandfather as the latest of six generations of machine makers working at the same premises.

Massive changes have occurred over the years. In the mid-Victorian era John Fletcher employed over fifty workers - carpenters, engineers, blacksmiths and millwrights - at Holly House and the Ironworks, along with his threshing gangs going around the farms. These men were capable of undertaking any job. Wood was the major material used, with the forge being used to shape and form metal components which were held in place by rivets and bolts. In modern construction no wood is used and metal is cut and shaped using a Messer Griesham CNC 'Magic Eye' gas profile cutter and then fixed in place by metal inert gas welding.

The workforce has declined steadily from five or six just after the Second World War until today. Now a very small team concentrates on the manufacture of specialised vee-framed sub-soiler/cultivators and equipment suited to the 250-plus horse-power tractors used by modern day farmers. They undertake high quality work for pig and chicken production units and in addition fabricate heavy-duty buckets for excavators and modify equipment to meet operators' special

Fig 92
Cambridge roller designed and built by T & J Fletcher, c1985.

Fig 93
Brice Baker grain bins and drying equipment, designed and installed by Fletchers, c1975.

requirements, for example, for demolition and furnace wrecking. After over 165 years' continuous service to customers, the Fletcher family is looking forward to many more years trading from Newport Ironworks.

Notes

[1] Local newspaper (unidentified), 1852

[2] *Lincolnshire Chronicle*, 28 August 1868

[3] Birch, N, 'Clayton & Shuttleworth & Co - Early Successes and a Strike!' in *Lincolnshire Past & Present*, No 50, (SLHA, 2002/3)

[4] John Fletcher, illustrated catalogue, 1872

[5] Implement catalogues, Lincolnshire Agricultural Society shows, 1869 et seq.

[6] Implement catalogue, Lincolnshire Agricultural Society show, Brigg, 1911

[7] John Fletcher senior, market book, 1880/81

[8] Indenture for deed of sale for Newport Ironworks between Thomas Marshall and John Fletcher, 13 May 1872

[9] Fowler, Mary, 'Between Trent and Ancholme' in *In and Around an Old-Fashioned Garden*, Jackson and Sons, Brigg, 1908

[10] Local newspaper (unidentified), 1891

[11] Thomas Fletcher's ledger, c1905

[12] Olney, R J, *Rural Society and Local Government in Nineteenth Century Lincolnshire*, (SLHA, 1979)

[13] T Fletcher and Sons, wages sheets for Newport Ironworks, 1904-11

CHAPTER 7

JAMES HART & SON OF BRIGG

IMPLEMENT & MACHINE MAKERS

Chris Page

The Rise and Fall of the Business

James Hart (1791-1848) established his business as a millwright some time before 1813. In that year the farm accounts of T J Dixon of Holton le Moor record Hart carrying out work on farm machinery. By March 1815 Hart was in partnership with William Berridge.[1] They both came from the village of Broughton, to the west of Brigg, and may have begun their careers as apprentices to Clarkson Pape, a millwright operating from Silversides in the early part of the nineteenth-century, Pape being one of the witnesses to James's marriage in 1812.[2]

Little is known about the Hart and Berridge enterprise, but, on 22 July 1824, at about 11.00 pm, this partnership was abruptly ended when William Berridge was fatally shot through the head whilst walking back to Brigg from Wrawby.[3] Robbery was the motive for this murder and his assailant was soon caught. James Wetherell, an unemployed chimney sweep from Brigg, was convicted of the murder and later hanged at Lincoln Castle.[4] Immediately following the death of Berridge, James Hart took his widow Ann as a business partner, giving notice on 27 August that the firm would continue under the name of Hart and Berridge. This arrangement only lasted a few months and in May 1825 the partnership was formally dissolved 'by mutual consent', with James continuing the business under his own name.[6]

In White's *Directory of Lincolnshire* for 1826 James was listed as working from Bridge Street in Brigg.[7] These were the details which appeared in subsequent trade directories for the town until Slater's *Directory* for 1849, when the business was under the title of Hart and Son. James had by then been joined by his eldest son, William (1818 to 1898)[8], although in fact James had died on 7 October 1848, aged 58.[9] At the time of the census of 1851, William was shown to be in sole control of the company, but he continued to trade under the name of James Hart and Son until 1872, when White's *Directory* listed the company under William's own name.[10]

On 1 October 1851, William Hart signed a 40-year lease with the Earl of Yarborough, who owned the land occupied by Hart's Ancholme Foundry. The lease began on the 13 March that year at a rent of £19 per annum.[11]

This probably marked a significant restructuring of the Yarborough estate in Bridge Street. From a

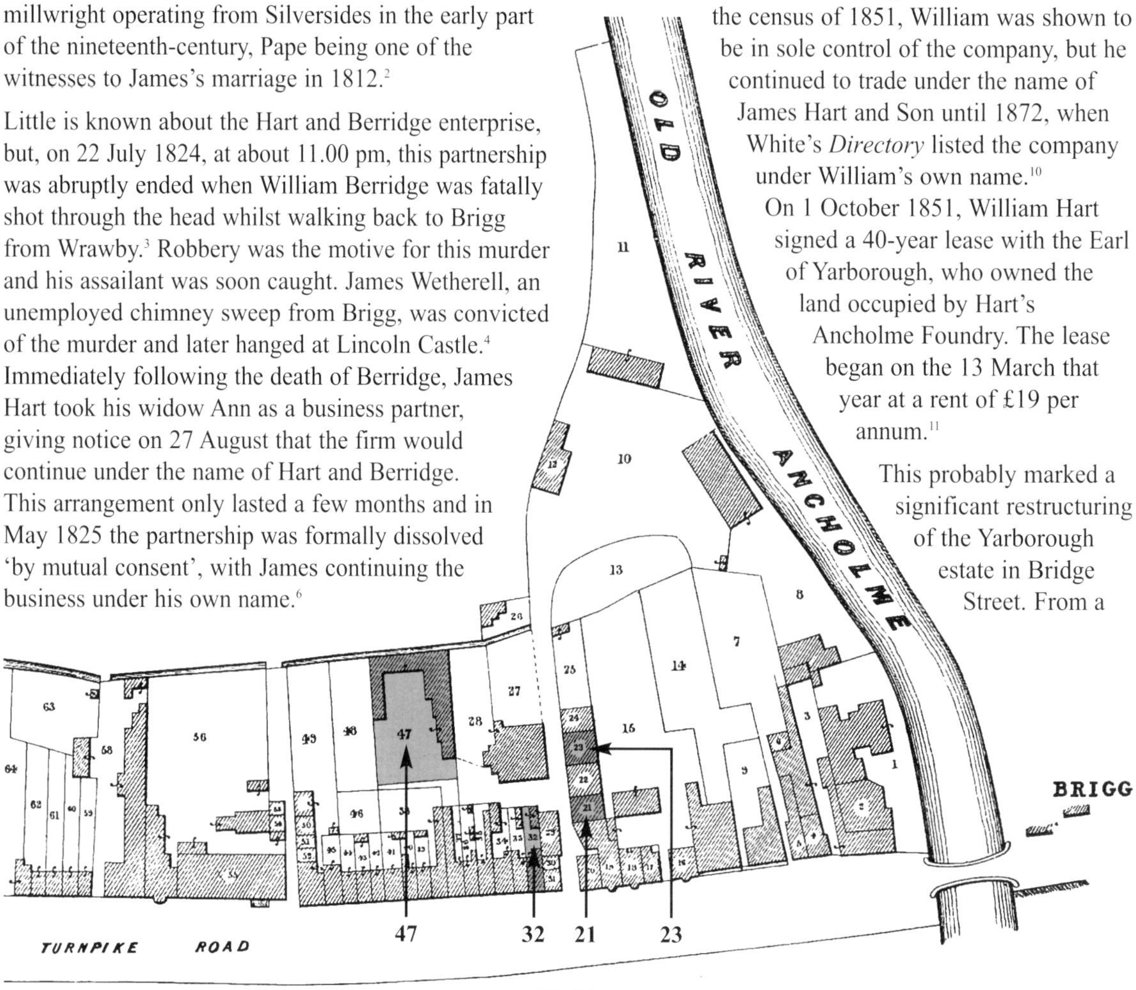

Fig 94
James Hart's property in 1842. 21: shop; 23: timber warehouse; 32: house and yard; 47: foundry and yard.

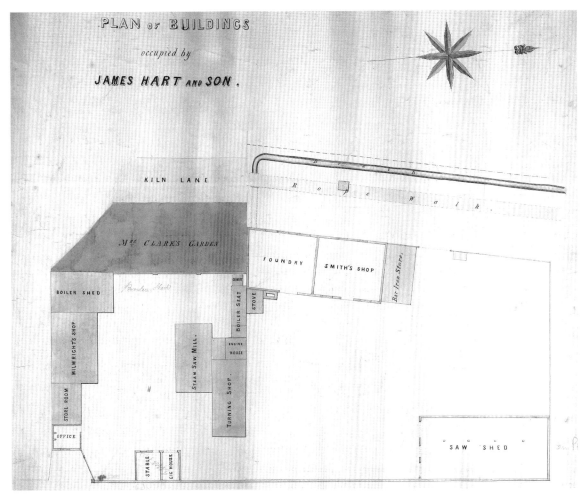

Fig 95
James Hart's foundry, c1850

plan of 1842, recording the tithe apportionment of the area *(Fig 94)*, Hart's foundry had a very different arrangement to that of 1851. The plan shows that there was a more dispersed nature to James's property at that time, with a timber warehouse and a shop (workshop?) as well as his house and gardens, which were separate from the main foundry buildings.

The plan and tithe records further show that James held the tenancy of Mill Carr, which was an area of grassland north of the town, four acres in extent, needed to support the horses required for haulage and personal transport.[12] The extensive layout of the buildings on this plan suggests that it had been established some years earlier and may well have been the premises occupied by Hart and Berridge in the 1820s.

Farming was entering a period of prosperity by the late 1840s, with increased investment by farmers, indicated by the term 'High Farming' of the 1850s and 1860s. This depended on the adoption of new technology and resulted in the use of new feeding practices, fertilizers and machinery.

Mechanisation of the farms gave immense opportunities to firms such as James Hart and Son. Between 1835 and 1841 business was developing to such an extent that a branch workshop was opened on the Grimsby Road in Caistor.[13] This branch continued to operate throughout the life of the company until c1868, with the directories listing Hart's business there as machine maker or agricultural implement maker. They also identified the firm under the name of William Hart, which may suggest that William initially managed this branch while his father controlled the Brigg Foundry.[14]

The census returns for Brigg provide some indication of how business at the foundry developed. In 1851

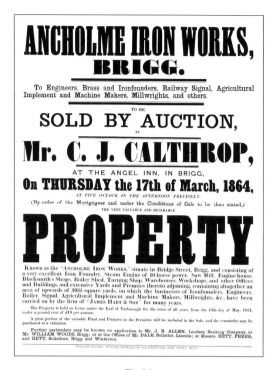

Fig 96
Sale poster, 1864, Hart's first attempt to sell the business.

it was employing 41 people, but by 1861 work had expanded to such an extent that it could support 60 men and 6 boys, making it the largest employer in the town.[15] A number of these men were skilled workers, such as brass moulders and boilermakers who had learned their trades in the industrial towns of the Midlands and Yorkshire. An interesting insight into the wage bill of this company came in 1855, when the people of Brigg donated money to the 'Patriotic Fund' in support of the British troops fighting in the Crimea. William Hart gave £3 to this fund and his workforce donated a day's wage each, which amounted to the sum of £9.10s.7d.[16] William Hart was also using the pages of the *Stamford Mercury* to attract additional, experienced men. In September 1854 he announced that he was 'in immediate want of two or three millwrights. Young men would be preferred.'[17]

As on the farms, investment was essential to the development of these small engineering companies, and for that purpose William borrowed money from the Lincoln and Lindsey Bank. However, by 1857 he had become indebted to them for 'a considerable sum of money on his current account.'[18]

To allow William to keep trading and explore new markets, the bank entered into a mortgage agreement on 16 October 1857.[19] This was secured upon the equipment in the foundry, and, with the Earl of Yarborough's agreement, consigned the remaining period of the lease to the bank for a term of 3 years.[20] William was now allowed to extend his account, providing it did not exceed £1,000, with a bank interest of 5%. A likely use of this money was to develop new product lines, with different designs of portable steam engines being produced. Also William began to explore new markets offered by the growth in the country's railway network, and in Kelly's *Directory* of 1855 he was listed as a manufacturer of railway signals.[21]

The fortunes of the company did not repay the earlier optimism and by 1861 Hart found he was unable to repay the bank, which subsequently foreclosed to retrieve as much funds as it could. William Hart was declared bankrupt by Thomas Freer, his solicitor, on 11 April 1864, using the Bankruptcy Act of 1861.[22] Under this agreement he was able to pay his creditors only five shillings in the pound, and to secure additional funds the remaining lease on the foundry, as well as its contents, were put up for sale. The sale was held on 17 March 1864 at the Angel Inn, Brigg. Mr C J Calthrop was the auctioneer and the sale was advertised nationally in the *Engineer* as well as many local newspapers *(Fig 96)*. A reserve had been set at £500, but there seem to have been only two bidders, a Mr Fowler and Mr Charles Smith; the latter was successful at £320.[23] Smith was acting as an agent for Sarah Walker and William's brother, Charles Frederick Hart, now a civil engineer in Devizes, Wiltshire. These two became joint owners of the business, allowing William to continue operating the foundry along with another brother, Adam Clark Hart. Later, in 1866, Charles Hart paid Sarah back her part of the investment, amounting to £160, to become the sole owner.[24]

Trade still did not improve and it was two years later, on 2 June 1868, that William and Adam Hart were again declared bankrupt, assigning their estate over to Joseph Parker, a builder in Brigg, and Edward Hudson Smith, who was an ironmonger in the town.[25] For a second time the Ancholme Foundry was put up for sale and an auction was held on 23 July 1868 at the Angel Inn, Brigg.[26] It seems that Charles Hart came to the rescue again, paying the Earl of Yarborough £400 for the premises. There then followed a further attempt to revive the business with William trading now under his own name. The title of the works changed from the Ancholme Iron Works to the Ancholme Foundry, although this name change may have happened a few years prior to 1868.[27]

The business, however, was not recovering and it came to a point when all parties did not wish to continue. William Hart now put the lease and equipment up for sale himself, advertising the foundry over a two week period in February 1872.²⁸ The buyer was Charles Louis Hett, a younger son of the prominent Brigg lawyer John Hett, whose law firm were Hart's solicitors. Hart finally surrendered the lease back to the Earl of Yarborough on 11 April 1872, so ending his relationship with the Ancholme Foundry.²⁹

The Steam Threshing Machine

The range of products James Hart made at the beginning of his business was typical of a country millwright and iron founder, with the building and repair of the many wind and watermills in the area providing a regular income. By 1826 James Hart was manufacturing corn threshing machinery and he supplied such a machine to one of T J Dixon's farms at Holton le Moor in September 1827.³⁰ The price of £13.3s.6d was paid on 31 January 1828, with a further 21 shillings to cover the board and lodgings for one of Hart's men working on the machine. Dixon's farm ledger book, which covers the years 1828 to 1831, records repairs and maintenance carried out by Hart on threshing and other machinery on the farm.

The records of the North Lincolnshire Agricultural Society give an insight into the range of items made by the firm from the 1830s:

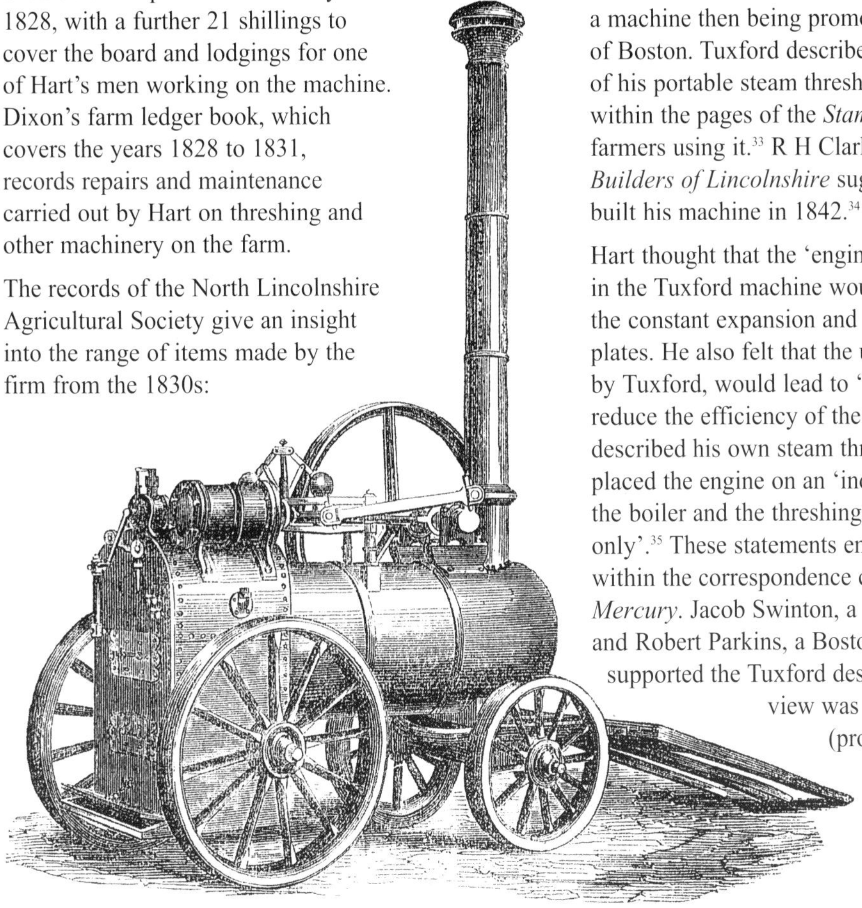

1837: blowing machine, dressing machine, turnip cutter, cast-iron land roller; 1843: scotch harrows; 1856: clod crusher, corn blower, roller, horse rake, dressing machine, swath rake, bean crusher, lawn mower, hicking barrow, wagon jack; 1858: seed drill; 1859: garden furniture; 1861: agent for eighty-four items, all manufactured by others.³¹

But the direction in which James was trying to expand his business became apparent on 8 December 1843, when he announced the completion of a portable steam threshing machine. He had put this machine to use on one of T J Dixon's farms at Holton, where it operated to their 'entire satisfaction'.³² James boasted that his new machine was both compact and substantial, since other portable steam engines he had seen 'have been imperfect in their arrangement'. He felt that his design was an important improvement, enabling it to thresh a great quantity of corn at 'an unprecedented small cost of fuel'. From the description it seems to have had its boiler, engine and threshing machinery all 'unified on one mobile frame'. This was in direct competition with a machine then being promoted by Tuxford and Sons of Boston. Tuxford described in detail the advantages of his portable steam threshing machine and listed within the pages of the *Stamford Mercury* some of the farmers using it.³³ R H Clark in his *Steam Engine Builders of Lincolnshire* suggests that Tuxford first built his machine in 1842.³⁴

Hart thought that the 'engine screwed onto the boiler' in the Tuxford machine would create instability due to the constant expansion and contraction of the metal plates. He also felt that the use of bevel gears, favoured by Tuxford, would lead to 'much friction' and so reduce the efficiency of the machine. James then described his own steam threshing machine, which placed the engine on an 'independent frame between the boiler and the threshing part, and [used] spur gears only'.³⁵ These statements engendered a lively debate within the correspondence columns of the *Stamford Mercury*. Jacob Swinton, a farmer from Billinghay, and Robert Parkins, a Boston engineer, strongly supported the Tuxford design, whilst an opposing view was expressed by 'Observer' (probably Hart himself writing under a pseudonym).³⁶

There is no record of how many threshing machines were made by the Harts, but they developed more than

Fig 97
Portable steam engine by James Hart, 1868.

one design. In 1854 William (James having died six years earlier) exhibited a machine of a more conventional pattern at the Royal Agricultural Society of England's annual show. This show was held in Lincoln, and was a combined event with the Lincolnshire and the North Lincolnshire Agricultural Societies, covering fields beside the Canwick Road coming out of the city. William Hart took this opportunity to present his latest portable steam engine along with a combined threshing and winnowing machine. The competition for an efficient threshing machine was intense. Each machine was rigorously tested, first with 100 sheaves of wheat, then with 100 sheaves of barley. The crop had to be cleanly threshed, the straw shaken and the crop winnowed ready for the finishing/dressing machines. Hart's threshing machine was rated at 6 nominal horse-power and took 9 minutes 54 seconds to process the wheat sheaves.

The judges found that it was reasonable in threshing the grain but rather less efficient in shaking the straw free of grain, with some grain and straw being broken. But it was found efficient in having the chavings and chaff free from corn. Out of the fourteen machines presented to the judges, Hart's thresher took a creditable ninth place, winning 69 points out of 85. (The winning machine, by Clayton and Shuttleworth, gained 84 points). In the trial with barley Hart achieved 71 points out of 85 and came fifth out of seven machines entered. Tuxford's machine achieved a higher result of 79 points for threshing wheat, but a lower score of 63 points for their barley sample. Tuxford's threshing machine sold for £100 against £95 for Hart's machine, but Hart's price was the same as that of Clayton and Shuttleworth's, which won the competition.

If this threshing display was only partially successful, the demonstration of Hart's portable steam engine was positively disastrous. This engine seems to have been a machine of a conventional design, rated at 8 hp. There were thirteen engines taking part in the working demonstration, but three failed to complete the trial, Hart's machine being one. It had begun the demonstration well, but then had to stop due to the 'consequence of imperfections in [the] driving pulley'. These 'imperfections' involved the fly-wheel being 'so badly adjusted' that it did not run true and would not keep the drive belt on. The result was that 'after several attempts the engine was removed' from the competition. With such a problem, had the engine remained at working speed, this misalignment could have created such a vibration that it would have

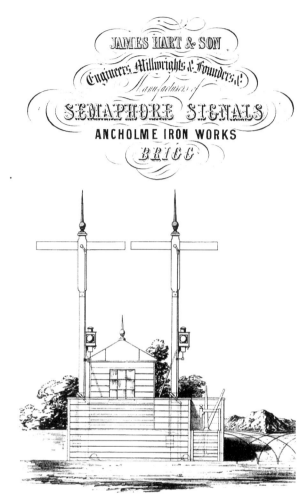

Fig 98
Railway semaphore signals and lamps made by James Hart in 1850s.

seriously damaged the engine and been unsafe to those close by![137]

Hart's machine and that shown by Hornsby were the only 8 hp engines presented. Of the others, three were rated at 7 hp, five at 6 hp, one at 5 hp, and the Tuxford was at 4 hp.[38] Hart's portable, however, cost £225, and was one of the cheapest at the trial for the given horse-power, apart from a 7 hp machine exhibited by Penistan of Lincoln, which cost £175. For Hart to have such a problem with his engine must have been a severe embarrassment and would not have aided his business development. The winning machine, of 8 hp, was made by Richard Hornsby and Son, Grantham, and cost £255, whilst Tuxford's 4 hp machine cost £190.

Other Products of the Ancholme Foundry

In an attempt to counter the bad publicity gained by being disqualified at the Lincoln Show, William ran an extensive advertisement in the *Stamford Mercury* for over sixteen weeks into January the following year. In it he promoted his portable engines, 'in which boilers of an improved principle are introduced, being constructed with flues and water spaces. The flues are arranged so that the flames traverse three times the length of the boiler, thereby realising the advantages of increased power, economy and durability of the engine.' His 8 hp portable steam engine, which weighed 62 cwt and used about 8 cwt of coal in a 10-hour day, was available, with governors, for £225. A smaller 6 hp machine, selling for £200, weighed 58 cwt and used 7 cwt of coal per day.

William's advert also announced the production of fixed steam engines 'of every description' as well as combined threshing machines. He stated that he was agent for the Oxfordshire firm of Samuelson, selling their digging machines and Cambridge patent rollers, of which Hart had provided his own 'improved heads and loose journals'. He sold Ransomes and Sims ploughs, Avery weighing machines, whilst 'all other farm implements were kept on sale', many being detailed in a catalogue available from the Foundry.[39]

Later advertisements expanded on the range of products Hart made and his role as an agent for other manufacturers increased. In White's *Directory* for 1868 William took a full page advertisement which included an illustration of one of his portable engines *(Fig 97)*. He also listed the range of equipment then being made at the Foundry, which included barn machinery, water tanks, heating systems and fencing. Also his agencies included Hornsby, Howard and Bentall among others.[40] This list of products was repeated in a later, full page advertisement which appeared in the 1872 issue of White's *Directory*.[41]

General engineering and foundry work seem to have formed an important part of Hart's business. An indication of this business is illustrated by the construction of a large, riveted sheet-iron water tank holding around 30 tons of water, made for the Brigg Water Works. This tank, completed in 1855, was placed on a 6-foot diameter stone and brick tower, 35 feet tall, in the centre of Brigg.[42]

By the 1860s Hart was becoming recognised for his water pumps, developing a range of hydraulic rams for use on farms and country houses. A letter from William Hall junior, of Redbourne, reported that 'the ram, erected by … Mr Hart, for supplying my house and farmstead with water, has given me the greatest satisfaction'.[43]

Hart was also said to have supplied the entire mill-gearing for Otter's Mill in Bridge Street, Brigg.[44] Perhaps the most interesting and intriguing area of production was the manufacture of railway signalling and railway lamps. This certainly showed enterprise, as this was a period of a rapidly expanding railway network within the British Isles. Development of these products seems to have begun around 1855, when Kelly's *Directory* for that year listed Hart as manufacturers of 'semaphore signals'.[45] To promote this work William commissioned high quality lithographic drawings from the firm of Waterlow and Son, London, showing six views of his products in trackside locations *(Fig 98)*.

This print must have been issued some time into the production period as it contained a list of railways

Fig 99
Cast-iron tobacco box said to have been made by James Hart & Son. It was presented to Scunthorpe Museum by the Long family.

using their signals, including the Manchester, Sheffield and Lincolnshire, and the London and North Western Railways.[46] On the lease agreement plans, possibly dating from 1857, pencil alterations show a lamp shop at the back of the premises. Also, when the Foundry dispersal sale was held in July 1868, there was still a quantity of 'improved' lamps remaining.[47]

Fig 100
William Hart's gravestone, St Mary's Churchyard, Barnetby.

Only one object made at the Ancholme Foundry in Hart's time is known to have survived. This is a cast iron tobacco box *(Fig 99)* held in the collections of Scunthorpe Museum and donated in the 1950s by Mr R G Long, who lived at the tower windmill in Scunthorpe. He identified it as having been made by Hart, but gave no further information. His father, Uriah Long, had replaced the post mill with a tower mill in 1858, which suggests that Hart may have been the millwright responsible for the mechanics and that this tobacco box was in commemoration of the new mill's completion.[48]

The foundry and workshops had grown over the years to reflect the changes in the firm's business requirements. The earliest detailed plan of the foundry is shown on the Broughton-by-Brigg tithe apportionment award of 1842 *(Fig 94)*. This dispersed layout was later consolidated and is found on a lease agreement with the Earl of Yarborough in 1851, when 41 people were employed.[49] As the work increased so the buildings were added to, and another plan, possibly dating from 1857, contains pencil additions of further buildings; by 1861 the workforce at the foundry had reached 60 men and 6 boys.[50] The use of these buildings was identified in an undated plan, possibly from the late 1850s, which shows the use of each structure, but does not include the lamp shop and wood store shown in pencil on the 1857 lease plan.

However, this final plan does seem to tie in with the description of the works seen at the 1864 and 1868 sales. This included a total area of 2.54 ha comprising: iron foundry, steam engine of 10 hp, saw mill, engine house, blacksmith's shop, boiler shed, turning shop, warehouses, workshops and other offices and buildings, and extensive yards.[51]

The effects of the works were sold separately and they included a 12 hp beam engine, powerful lathes, platform weighing machines, a Porter & Co No. 2 'National' gas apparatus, as well as 20-dozen large railway lamps, 'some on a new improved principle', and agricultural implements.[52] This was how the works looked at the end of its days under William Hart, who by 1871 employed just 14 men and 2 boys.[53] It seems that the site's sale in February 1872 was by the mutual agreement of all parties, formalised by the surrender of the lease on the 11 April.

William, who was living in a house held on a yearly lease adjoining the works in Bridge Street, then moved to 21 Bigby Street, Brigg. There he continued to trade in a reduced way, the census of 1881 recording him as an engineer and a millwright. In 1891 he described himself as a millwright and engine maker. However, it is unlikely he was making engines, but was possibly repairing such machines and occasionally, using the pages of the *Stamford Mercury*, to sell them, as in October 1872 when he offered a second-hand 8 hp portable steam engine for sale.[54] Larger equipment was passing through his hands, as recorded in March 1873 when he advertised a set of steam cultivating machinery, complete with 400 yards of steel rope, 1,500 yards of manila driving rope, anchors, porters and a Fowler four-furrow plough and a seven-tine cultivator, 'the above equal to new'.[55] William Hart died in 1898 and is buried with his wife Elizabeth at Barnetby old church yard of St Mary's *(Fig 100)*.

James Hart's career had begun as a small country millwright, but he showed that he had aspirations far above many of his rivals.[56] He was prepared to embrace the new mechanical ideas and to experiment, which is illustrated by his work with threshing machines and steam engines in the 1840s. James and his son William were also not averse to exploring more unusual new market opportunities such as the development of railway signalling. As a result, the business flourished for a while, becoming one of the largest employers in Brigg by the 1860s, and also sustaining a branch workshop in Caistor for over 30 years. Yet the business ultimately failed financially.

In the firm's early years it had held a virtual monopoly for general engineering within the area. There were other millwrights operating, but their businesses were small in comparison. The Harts' only major rival was Isaac Spight, who had established his 'Victoria Iron Works' across the road from the Ancholme Foundry by the 1840s. By the end of the nineteenth century, Spight had developed a sizeable business which manufactured agricultural machinery. However, Hart was well established by the 1840s, having been able to grow as the local economy grew. He was able to supply the town's expanding industries and to benefit from the extensive mechanisation of the farms in the 1850s and 1860s.

The full reason for his failure to survive and expand is not clear, but it would seem to be in part through lack of capital. It may also be involved with technical ability, as engineering was becoming more competitive and sophisticated. The failure of the portable steam engine at the Lincoln Royal Show of 1854 seems to have been a crucial turning point in confidence and his fortune. With the larger county machinery manufacturers such as Hornsby, Foster, Marshall and Clayton rapidly expanding, let alone the national companies such as Ransomes of Ipswich and Howard of Bedford, the exclusive market that the Harts once enjoyed was becoming swamped. With the loss of support from their bank, the end was inevitable, leading to the eventual demise of the business.

Notes

[1] Lincolnshire Archives (LAO), DIXON 9/1/13/72; *London Gazette* 4 March 1815, Messrs Hart and Berridge, millwrights of Brigg, are creditors of John Croft, a miller in Caistor; Pigot, *Directory,* 1822-1823.

[2] Clarkson Pape (c1765-1846); *Lincolnshire Rutland and Stamford Mercury* (LRSM), 4 June 1813 p.3 col. 3. Pape advertised for 4 or 5 journeymen millwrights; Broughton Parish Marriage Registers, 1812.

[3] *LRSM*, 30 July 1824, p.3, cols 2 and 3.

[4] *LRSM*, 6 August 1824, p.2, cols 4 and 5; 13 August 1824, p.3, col. 2.

[5] *LRSM*, 27 August 1824, p.3, col. 4.

[6] *LRSM*, 6 May 1825, p.3, col. 5.

[7] W. White, *Directory of Lincolnshire*, 1826.

[8] Slater, *Directory of Lincolnshire*, 1849.

[9] Census returns for Broughton by Brigg, 1841 and 1851; Broughton Parish Records: James Hart was buried at St Mary's Church, Broughton.

[10] W. White, *Directory of Lincolnshire*, 1872.

[11] Lincolnshire Archive Office, YARB 74.

[12] LAO, LRO I 291, Broughton by Brigg Tithe Award and Plan, 1842. Interestingly it shows that an immediate neighbour, the solicitor John Hett, and his son Charles Louis Hett, who later took over the foundry from William Hart in 1872, may have been born here.

[13] Pigot, *Directory of Lincolnshire*, 1835 and 1841.

[14] Hart may have been linked at this time with Thomas Kynman, a machine maker and millwright operating on Grimsby road in Caistor.

[15] They were large enough for their workers to have formed 'a capital brass band'.

[16] *LRSM*, 5 January 1855, p.3 col.8. If this figure is taken over a 6-day week it means Hart had a possible annual wage bill of £2,824.18s.0d in 1855.

[17] *LRSM*, 29 September 1854, p.3, col 6.

[18] LAO, YARB. 66.

[19] LAO, YARB. 69.

[20] LAO, YARB. 67.

[21] Kelly, *Directory*, 1855.

[22] *LRSM*, 22 April 1864, p.2, col. 5.

[23] LAO, YARB. 70 and 72.

[24] LAO, YARB. 68.

[25] *LRSM*, 12 June 1868, p.2, col. 4. A number of drain covers are cast with the name 'E H Smith, Brigg' and can still be found in Brigg and Winterton, as well as other villages in the area.

[26] LAO, YARB. 72; LRSM, 10 July 1868, p.7, col. 2.

[27] LAO, YARB. 68; YARB. 72; White, *Directory*, 1872, p.19.

[28] *LRSM*, 9 February 1872, p.1, col 7.

[29] LAO, YARB. 68.

[30] W. White, *Directory of Lincolnshire* 1826; LAO, 1-DIXON/22/4/1, p.117.

[31] *LRSM*, 8 December 1843 p.3, col.7.

[32] LAO, STUBBS 1/16/5.

[33] *LRSM*, 16 June 1843, p.3 col. 8 and 4 August 1843, p.3, col. 8.

[34] R H Clark, *Steam Engine Builders of Lincolnshire*, (SLHA reprint, Lincoln, 1998), pp.114-115.

[35] *LRSM*, 8 December 1843, p.3, col. 5.

[36] *LRSM*, 22 December 1843, p.4, col. 6; *LRSM* 29 December 1843, p.4, col. 7.

[37] *LRSM*, 21 July 1854, Supplement p.3, col. 2.

[38] *Royal Agricultural Society of England Journal*, 1854, p.374.

[39] *LRSM*, 5 January 1855, p.3, col. 8.

[40] W. White, *Directory of Lincolnshire*, 1868, p.529.

[41] W. White, *Directory of Lincolnshire*, 1872, p.19.

[42] *LRSM*, 9 December 1853, p.2, col 2; 24 August 1855, p.3, col.4 and F Henthorne, *The History of Brigg Grammar School Trustees*, (Boston, 1959), p.107.

[43] A testimonial letter sent to Charles Louis Hett, the successor to Hart at the Ancholme Foundry, on 18 August 1874, and published in Hett's 1877 catalogue. This was for a ram pump installed at Redbourne Hall in 1860.

[44] Henthorne *Brigg History*, p.97, *LRSM* 1869 and *LRSM* 15 May 1903, p.8, col. 6.

[45] E. R. Kelly, *Directory of Lincolnshire*, 1855.

[46] LAO, STUBBS 3/20.

[47] F Henthorne, *The History of Nineteenth Century Brigg*, p.97; LAO, YARB 75.

[48] Scunthorpe Museum collection number: 0058. There must be many more examples of this company's work, but as they do not seem to be identifiable they must remain anonymous!

[49] LAO, LRO I 290; LAO, YARB 74; 1841 census for Brigg.

[50] LAO, YARB 75; 1861 census for Brigg.

[51] *LRSM,* 4 March 1864, p.7, col. 5; *LRSM,* 10 July 1868, p.7, col. 2.

[52] Henthorne *Brigg History*, p.97. (Porter & Co manufactured gas equipment in Lincoln.)

[53] 1871 census for Brigg.

[54] *LRSM*, 18 October 1872, p.1, col. 6.

[55] *LRSM*, 21 March 1873, p.1, col. 6. On the same page C L Hett had put the house adjoining the offices of the Ancholme Foundry up for rent, which may have been Hart's family home.

[56] Both James Hart and William Berridge were Methodists; James was a trustee of the Wrawby Street Chapel in Brigg in 1839; *The Register of Persons Entitled to Vote - Parts of Lindsey 1839*, p.150, Glanford Brigg, No 9499.

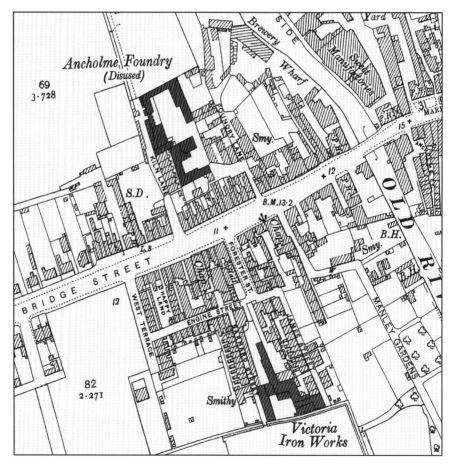

Fig 101
1907 OS map of Bridge Street area, Brigg. The Ancholme Foundry was occupied by James Hart (c1820-1872), C L Hett (1872-1895) and Peacock & Binnington (1896 to present). Isaac Spight occupied the Victoria Iron Works (c1840-1896), followed by J B Edlington (1896-1979).

CHAPTER 8

C L HETT OF BRIGG

HYDRAULIC ENGINEER

Chris Page

The Man and His Career

Charles Louis Hett was born in 1845, one of thirteen children, to John and Louisa Hett. His father had become one of the most prominent and influential men in Brigg, holding a partnership in a long-established firm of solicitors: Hett, Freer and Hett. At the time of his death in 1878, John Hett held many of the key administrative roles within the area, including Clerk to both the Commissioners of Ancholme Drainage and Navigation and the North Lincolnshire Agricultural Society.

Charles received a private education in Worksop, but did not take up law, unlike his elder brothers, Roslin and John junior. He was more practically minded, leaving the town to train as an engineer with the firm of Benjamin Hick and Son, in Bolton, Lancashire.[1] Following this, it is known through the 1871 census that he lodged in Ripley, Yorkshire, and worked as a mechanical engineer. Shortly afterwards he returned to Brigg with the intention of establishing his own engineering firm, and in April 1872 Charles, then aged 27, took over the Ancholme Foundry from William Hart.[2]

The Ancholme Foundry had been established in Brigg around 1813 by the millwright James Hart.[3] Hart's business had grown to become one of the town's largest employers by the 1850s, but by the 1860s it suffered from financial difficulties.[4] The foundry was then occupying an area of 2.54ha and was run by James's son, William. The works not only included a foundry, but also a forge, boiler and turning workshops, making it an ideal location for Charles Hett to set up his enterprise.[5] From the beginning, Hett started to develop the general engineering opportunities of the business rather than rely upon the agricultural industry, as his predecessor at the foundry had done.

A Lincolnshire trade directory of 1876 described Hett as an engineer, millwright and agricultural implement maker.[6] Charles appears in the 1881 census returns at 12 Bigby Street, Brigg, along with a housekeeper and maid. This was just nine houses away from William Hart, who was then living at 21 Bigby Street.[7] Ten years later, in the 1891 census, Charles was occupying what was then called Foundry House in Bridge Street. He was unmarried, describing himself as a 'turbine and

Fig 102
Snitterby Carr bridge over the Ancholme Navigation, built by Hett in 1872, one of his earliest projects (photograph 2005).

centrifugal pump manufacturer'.[8] This corresponded to a change in his listings in trade directories, *Kelly's Directory* for 1892 identifying him as a 'hydraulic engineer', and adding in 1896 that he made 'water power and rural water supply a speciality'.[9]

His interest in water management is illustrated by a number of papers published in the *Minutes of the Proceedings of the Institute of Civil Engineers*. Hett had joined the Institute in 1877 as an Associate Member and remained with them until resigning in 1895.[10] He published a number of technical books in the 1880s, including a discussion on rural water supply which appeared in 1888 and a table of power for belting and shafts in 1889.[11] In addition to these, he wrote articles for the *Engineering News* in 1884 on rural water supply.[12] His product catalogue, copies of which exist for 1887 and 1889, was more than just a description of the models he produced; he included technical discussions designed to aid the prospective client in their choice of machine.[13] The business seems to have thrived, but there is no indication of the size and changes within his labour force to enable a clear picture to be formed. However, one small indication of the scale of the firm's activities comes from the complaint made by neighbours of the foundry in 1892. They complained about the amount of smoke and demanded that the main chimney be raised by 6 feet to reduce the nuisance.[14]

In 1895, when Hett was 49 years old, he withdrew from the business, selling off the factory and plant. He had married Hannah Foyster in that year; she had been born at Pendleton, Lancashire, close to where Charles originally trained, and they were now living in Wrawby Street, Brigg.[15] By the next census, in 1901, Charles had moved into retirement at 'Springfield', a large house in Wrawby that had originally been built by his relative, the solicitor Thomas Freer. Charles described himself at the time of this census as a retired mechanical engineer, and he busied himself in a small workshop where, among other projects, he developed optical surveying equipment, taking out at least two patents on his inventions by 1899.[16]

In addition to his engineering interests, Charles studied birdsong, of which he was particularly knowledgeable. He published his findings in 1898 as the *'Dictionary of Birdnotes'*, printed by Jacksons in the Market Place, Brigg. They were the printers that had produced the majority of Hett's catalogues and text books. Hett was also a founder member of the Zoological Photographic Society, established in London in 1899, becoming its first honorary secretary until 1908.[17] Charles Louis Hett died from heart failure on 22 September 1911, aged 65, and is buried in the family grave within Brigg town cemetery.[18]

The Early Years: Agriculture, Engineering and Steam

Charles ran two businesses from the foundry site, the first being the engineering enterprise, the second a farm contracting service. This latter activity ran under the name of the Brigg Steam Cultivating Company, which had been established by July 1872.
The published prospectus stated that they proposed to spend about £4,000 on buying two sets of steam cultivators 'direct from the makers, not through agents, so saving on commission' and Hett was listed as their secretary.[19] They reported that they had the support of many local principal farmers, and the enterprise seems to have been successful, for it appears in the trade directories from 1876 to 1889.[20]

The training Charles received with Benjamin Hick and Son (later Hick, Hargreaves & Co) would have been very solid, as they were well known for producing

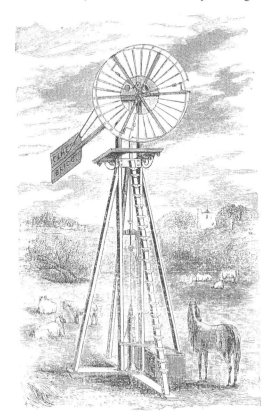

Fig 103
Hett's patent self-regulating wind pump. It was exhibited at the Lincolnshire Agricultural Society Show in 1880.

Fig 104
Overshot waterwheel pumping engine at Branston, Lincoln, made by Hett (photograph c1993).

some of the largest steam mill engines and machinery in the country. The production of water turbines also formed part of their business. After he left Bolton in 1868, he moved back to Brigg (via a spell in Yorkshire) and soon resumed links with the area. His subscription of 10 shillings to the North Lincolnshire Agricultural Society in 1871 is but one example.[21]

One of Hett's first large contracts was for the Snitterby Carr accommodation bridge on the Ancholme Navigation *(Fig 102)*. Tenders were required to be submitted to the Navigation Trust by 26 March 1872 and work had to begin quickly because the bridge was required to be completed that November. Charles may have had access to the foundry prior to William Hart relinquishing the lease, which enabled him to begin this work. His quotation of £335 had not been as low as a London firm of engineers, Thomas Head (£325), but he was well below that of Dossor and Weddall of Grimsby, which required £1,500 to complete the work! Hett's quotation was accepted and the 19.8 m span was finished on 18 November 1872.[22] The bridge still stands (2006) and proudly bears an oval cast-iron plaque in the centre stating that it was made at the 'Ancholme Foundry 1872'.

Hett's initial range of products seems to have been similar to those made by Hart, and, indeed, it is very likely that Hett would have used patterns and designs inherited from William Hart. For example, in 1874, Charles supplied all the machinery for the tower mill at Scotter.[23] In the same year the Governors of Brigg Grammar School looked to Hett for a new iron waterwheel for their corn mill at Fulsby (near Coningsby); however, his estimate of £135 proved to be too high.[24]

Such millwrighting activities were blended with agricultural work and a selection of his early products appeared on the stand of Edward H Smith at the 1873 Lincolnshire Agricultural Society (LAS) show at Gainsborough. E H Smith was a long established Brigg ironmonger and implement agent.[25] He had been involved with Hart and the Ancholme Foundry for many years and now he supported Charles Louis Hett. He included Hett's 'Long Boss' Cambridge cast iron roller, which was 7 ft 6 in wide by 2 ft diameter, selling at £13. There was also a mangold drill for £7.10s made by Hett, with separate travelling wheels at an extra 10s. A press drill was exhibited for £13, also cast-iron columns for agricultural buildings at £1 each.[26]

Charles Hett organised his own stand at the following year's show, which was held at Grantham. In his display he exhibited a turnip cutter and a root pulper, both manufactured by Hornsby, and also a chaff cutter made by the Sheffield firm of Crowley, which he continued to sell until 1876. This possibly indicates that Hett was now moving away from manufacturing agricultural equipment and towards being an agent for such machines. The centre of this display in Grantham was a 2 hp horizontal, high pressure, expansive steam engine of an 'improved construction', designed and made by Hett. It was powered by a vertical, tubular boiler also made by him. The engine cost £35, whilst the boiler was priced at £40.[27]

Later, in 1874, he developed this horizontal steam engine into his 'Universal' type, which he produced as a 2.5 hp model, increasing to an 18 hp machine, which had 14 in x 28 in cylinders. The *Engineering Journal* described it as a 'neat little horizontal engine' with many 'novel features'.[28] One of these engines was supplied to the Composition Works of Day and Osgerby at Brigg in the summer of 1875.[29]

The following year he supplied engines to St Denys's Sewage Works, which was part of the Portsmouth Drainage project in Southampton. Hett took a 2.5 hp engine to the 1875 LAS show at Grimsby and to the 1876 Royal Agricultural Society of England (RASE) show at Birmingham. Here the engine was described as 'a very compact and well made horizontal engine ... well designed for driving chaff cutters, root pulpers etc', which was possibly why he was selling other manufacturers' barn equipment.[30]

Ram Pumps and Chain Pumps

Production of the horizontal engine continued into the 1880s, and Hett exhibited examples at Lincoln in 1876 as well as the Brigg agricultural show of 1880. By 1876 Hett stopped producing his own boilers and used those made by Abbott & Co, boiler makers at Newark. Purchasing boilers may indicate that the space taken by the boiler makers at his works was required for producing new lines. His display at the Grimsby show of 1875 had included hydraulic rams for the first time. A number of two-inch diameter, high-lift rams were shown and a similar display was mounted at the following show in Lincoln.

At the 1880 Lincolnshire Show in Brigg there was a change of emphasis in the range of implements on display. This was his home show, so Charles presented a major exhibition of his equipment, which took pride of place by occupying stand No 1. His machinery even supplied the water to the showground, feeding the water troughs and tanks which he had supplied for the working machinery and the livestock exhibits. He also supplied water to an ornamental fountain, which again he had manufactured. It was the range of products that he now exhibited which indicated the changes in his business.

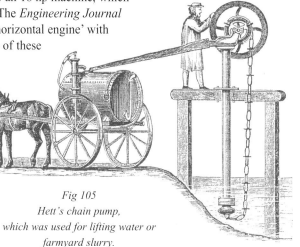

Fig 105
Hett's chain pump, which was used for lifting water or farmyard slurry.

As well as single- and double-cylinder horizontal steam engines, he first showed his 'self-regulating' wind pump *(Fig 103)*, which stood on a thirteen-foot high wooden tower. There was also a selection of hydraulic rams and water wheel pumps. This group of machinery illustrated Hett's move towards hydraulic engineering, especially in supplying waterworks to rural communities.

In 1876 he had shown a selection of his hydraulic rams at the Birmingham Royal Show to good reviews. However, the comments on his display at the 1878 Royal Show at Bristol go some way to explain why he was moving in that direction:

> This stand was quite a centre of attention, and numberless were the inquiries made with respect to the leading features of the double-acting pumping rams. This was scarcely matter for surprise. Large numbers of those who visited the "Royal" show occupy mansions, farmsteadings, or villas, or live in villages which are far away from any source of the great essential of health and comfort - water..... Visitors, while watching these improved hydraulic rams at work, saw at once the means of having water brought, at comparatively trifling cost, from almost any distance.[31]

The ram pump was known by the end of the eighteenth century and, after many new designs were developed in America in the 1850s, it began to gain a wider interest from the public. Hett responded to this market and had two 'Anglo-American' pumps in operation on his stand at Bristol, one having a glass air vessel to exhibit its operation. At Brigg he had nine such machines in operation, including one made by the Reading firm of Easton, Amos and Anderson. Hett's machines sold from £2.16s.0d for a small 1.25 in model to £15 for a 3 in improved high-lift ram.[32]

These ram pumps became a common feature of the countryside by the end of the nineteenth century, yet were rarely seen. It was their sound that identified them, as the action of the valves opening and closing produced a regular thump that travelled over some distance. The manufacture of rams at the Ancholme Foundry had begun under William Hart, as a testimonial letter sent to Hett from William Hall junior of Redbourne confirmed. In this letter, dated 18 August 1874, Hall praised the efficiency of Hart's pump, which he had purchased 14 years before. A catalogue published by Hett in 1877 implies that he had first brought out his high lift 'improved hydraulic ram' the year before, adding to the 'Anglo-American ram'. He tried to make these machines more accessible to possible buyers, publishing in 1875 a *Directory for Gauging Streams* to help people decide on the best model for their needs.

Alongside the ram pumps at the 1880 Brigg show he ran pumps powered by a waterwheel. One three-foot diameter waterwheel powered two two-inch pumps and was the standard model supplied by Hett for mansions and larger farms. This size cost £35 and was similar to one used at Hooton Levett Hall near Rotherham, lifting 1,700 gallons to a height of 50 m, through 488 m of piping every 24 hours.[33] A larger waterwheel, ten feet in diameter, powered pumps supplying a reservoir at Branston, near Lincoln in 1880.[34] This wheel and pump were still surviving in 2006 near the centre of the village *(Fig 104)*.[35]

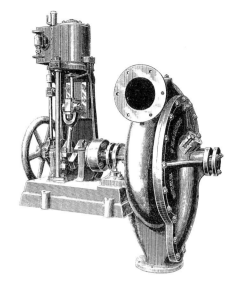

Fig 106
Centrifugal pump for drainage at Minden, Holland, supplied by Hett, 1880.

To supply the water carts at the Lincolnshire Agricultural Show at Brigg, Hett used another type of pump, a three-inch chain pump. This was a very simple machine involving a continuous chain with a series of discs or pistons that held the water as the chain rose up through a pipe from the well. The pump could also be used to lift liquid manure from tanks, and in an advertising supplement to the show catalogue Charles illustrated his pump, complete with a smocked farm worker *(Fig 105)*.

In the same supplement he included a drawing of his 'patent Noria or improved Persian wheel'. This was an improved chain pump with a series of cast iron buckets. An illustration of this pump appeared in Hett's advertisements of 1880, which were published in journals as well as technical farming books at the time.[36] Chain pumps became very popular and can still be found, though not in operation, on some farms today, but the bucket pumps were more limited in their use.

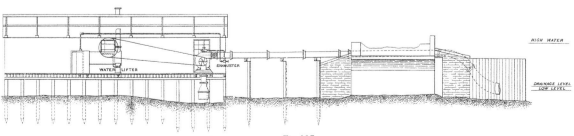

Fig 107
Section of drainage engine installed at Messingham, Lincolnshire. Hett supplied the centrifugal pumps.

This was not the case with the centrifugal pump, which had applications on farms, in industry and on board ship. Hett supplied this pump for all such locations and he published testimonials from satisfied customers, including the Hornsea Steam Brick and Tile Works, and the Tunnel Lime and Stone Works at Kirton Lindsey.[37] These pumps appeared in his catalogue for the first time in 1877, where he explained the advantages of his 'patent side-opening'. The *Implement and Machinery Review* of September 1880 gave a further explanation of this type of opening, which allowed access to the central impeller or disc 'by merely unscrewing a few nuts', enabling the machine to be serviced without disturbing any pipework.

Hett promoted this pump as being ideal for sewage and drainage work, for contractors, for emptying tan pits or clay pits and quarries, also irrigation and 'for sheep washing in the Colonies'.[38] By 1880 Hett had also introduced a vertical, direct-acting steam engine to connect with these pumps, especially for circulating water in surface condensers on ships, economising on the space available.[39] A similar pump was installed in the SS Eldorado and was described in the *Engineer*, 18 June 1886. This had a two-foot disk and gave a full discharge at a height of 5.0 m when it was running at 190 rpm.[38a]

Turbines and Larger Engineering Projects

Hett was turning increasingly towards the industrial market as depression bit deeper into agriculture. His policy of exhibiting seems to have moved away from local shows towards national and international events. For example, he had taken the opportunity to promote himself to a wider public at a number of the Royal Agricultural Society Shows. He also sent his turbines to the Amsterdam International Exhibition of 1884 and he gained two 'first orders of merit' for centrifugal pumps and pumping engines at an exhibition in Adelaide, Australia.[40]

He only exhibited twice more with the Lincolnshire Agricultural Society: at Gainsborough in 1883 and finally when the Society returned to Brigg in 1891. The Gainsborough stand allowed the public access to three sides of a large display of centrifugal pumps ranging from three-inch to twelve-inch diameter. The stand at Brigg enabled Charles to show his local town and the people of Lincolnshire again just what he could do.

Among the Hett drainage pumps installed in the 1880s was a 36-inch centrifugal model supplied from Brigg and installed at Minden, Holland *(Fig 106)*. It was driven by a steam engine supplied by the Dutch firm,

Fig 108
Overshot waterwheel pump with turbine developed by Hart by 1878. For smaller projects with a more limited supply of water, a ram pump was employed.

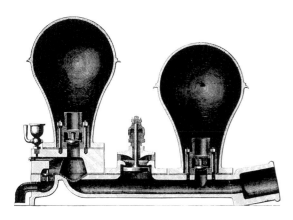

Fig 109
Hett's ram pump advertised as suitable for rural waterworks.

Machinefabrick of Breda. It operated at 120 rpm, discharging 1,700 gallons of water per minute to a height of 6 feet through ten-inch pipes. The cost of the complete installation was £250. A similar 'Accessible' centrifugal pump, powered by a double-cylinder, semi-portable steam engine, was installed on the Trent by Hett at Butterwick in March 1882 to drain 3,250 acres of land in the Messingham district of Lincolnshire *(Fig 107)*. It could discharge 10,000 gallons of water per minute through 21-inch cast-iron pipes. It was put to the test in the first winter of operation when the heaviest rainfalls known coincided with one of the highest recorded tide levels on the Trent. The pump was running for 197 consecutive hours, stopping only for lubrication, and proved to be a complete success, whilst the neighbouring scoop wheel pumping engines were completely drowned.[41]

About this time Hett also installed a similar pump for the Duke of St Albans in his carr land adjacent to the river Ancholme at Redbourne. He used a 14 hp double-cylinder, semi-portable steam engine, set in a brick building by the river, to power a centrifugal pump with 18-inch pipes to deliver 21.10 tons of water per minute. The cost of this engine and pump was £641. When that part of the Redbourne estate was sold in 1927 the pump was still operating, though by then powered by a 32 hp Crossley oil engine.[42]

By the early 1880s turbines were the main product coming out of the Ancholme Foundry. They had taken such a dominant role that by 1882 Charles had changed the name of the works to the Turbine Foundry. It had taken some time after acquiring the Foundry before Hett felt able to begin manufacturing his own designs of turbine. He discussed his ideas with James Emerson in Springfield, Massachusetts, one of the most experienced people working with turbines at the time and his reply survives in records at Gilbert Gilkes' archives in Kendal.[43] Emerson was disparaging of those using 'scientific methods' to design such machines, recommending them to follow the more practical manufacturers.[44] Hett wrote: 'When opportunity occurred, we carefully examined the various types of turbine, both of home and foreign manufacture. Our earliest turbines were of the Centre-Vent type…' This design followed the American practice as seen in the 'Francis' type of turbine. His experiences showed that a universal machine was not possible and so this led to various models, designed to meet specific requirements of the customer.

His first turbine was a modification of a type known as the 'Vortex Wheel', invented by Professor Thompson. This was the 'Centre-Vent' *(Fig 111)* introduced by 1880 and shown that year at the Brigg show, costing £60.[45] In the same year the *Implement and Machinery Review* for August contained an article on the turbines produced by Hett, referring to Blackwell Mill near Buxton, owned by the Midland Railway, where Charles had installed one of the first of his turbines. Other models soon followed, meeting the needs of a growing number of customers. The 'Trent' was a popular small turbine appearing around 1882.

Fig 110
The 'Hercules' turbine, the largest made by Hett.
(The photograph is thought to show Charles Louis Hett himself standing alongside).

Fig 111
Hett's first turbine, the 'Centre-Vent', c1880. It had a horizontal spindle for smaller power needs.

There then followed the 'Girard' for driving mills with variable water supplies, but Hett 'strongly recommended these… for driving dynamos direct from the turbine shaft'.[46] The 'Girard' was from a German design, but the 'Little Giant' had its origins in America. This was a small, cheap turbine often used to replace overshot waterwheels in corn mills.

A further two designs, which again had American origins, were built by the end of the 1880s. This included the 'Victor' and the 'Hercules' models, with the latter machine being made up to 54-inch diameter *(Fig 110)*. On Hett's stand at the 1891 Brigg show were a twelve-inch 'Little Giant' costing £35 and two 'Hercules' turbines with a fifteen-inch vertical shaft. A 'Victor' turbine, which he had illustrated in his catalogue of 1889, was also shown, but this machine was made for him by an American firm, 'Stillwell, Bierce and Co', costing £50.[47]

Also at the Brigg Show a 24-inch Pelton wheel *(Fig 112)* in a case was on display, costing £45. This was being built at the Foundry by 1889, using a bifurcated bucket to improve its efficiency. It was popular in the mining districts of California and Hett was one of the first manufacturers in Britain to make use of this design.[48] To supply a contract for Pelton wheels to mines in the Andes, in South America, Hett sectionalised them, enabling the equipment to be delivered to the mines by mule. It was a Brigg-built Pelton wheel that was supplied to the Lynton electric lighting station, North Devon, in 1895. This joined one of Hett's 'Little Giant' turbines which had been installed in 1889. The 14-inch turbine was rated at 200 hp at a working head of 95 ft.

By the time Charles Hett decided to retire and close the foundry he is said to have constructed 257 turbines.[49] One of these was located at Haverholme Lock, by the Sleaford Navigation, supplying water to Haverholme Priory, stables and home farm *(Fig 113)*. It still survives (2006) by the lock, but the large country house is now but a picturesque ruin.

He is also known to have supplied the turbines for pumping the sewage system in Sleaford. This was installed close to Cogglesford Mill, also on the Sleaford canal, and the foundation pit still survives. Another project was to power Caudwell's Mill, near Bakewell in Derbyshire, and one of his turbines was installed at the Sullatober Bleaching Company in Carrickfergus, County Antrim, Northern Ireland. However, Hett's turbines were mostly replaced by other, larger models early in the twentieth-century. In 1894 a 'Hercules' turbine, powering one of Hett's 6-inch 'Accessible' centrifugal pumps, was installed to drain a telegraph subway under the Manchester Ship Canal.[50] Regrettably, this too has not survived.

The contents of the Ancholme works were finally sold off on the 15 August 1895 by the auctioneers Wheatley Kirk, Price and Gaulty, who were based in London and Manchester. Their catalogue included:

> ….engines, planing, slotting, drilling, shaping, boring, grinding and other machines. Numerous ungeared single, double and treble geared lathes up to 88 in face plates and 28 ft beds, two 10 ton foundry cranes, 3 ton overhead traveller, 2 ton and 4 ton derrick cranes, loose plant and utensils, bins, foundry boxes and plant, office furniture, stores and numerous other effects.

> The new stock includes 11.5 in., 15.5 in. and 19.5 in. 'Little Giant' turbines, numerous 'Trent', 'Victor', 'Hercules' and other turbines; 5 in., 6 in., 10 in. and 12 in. centrifugal pumps etc.[51]

Fig 112
A Pelton wheel made by Hett in 1880s. It was favoured by mining engineers.

From this description it can be seen that the organisation of the Foundry had become far more extensive and specialised than during Hart's period. The sizes of the lathes and cranes show that they were able to cope with heavy, bulky parts and the range of machine tools indicates the degree of engineering skills the firm possessed for the production of high speed machinery, such as the turbine.

Fig 113
Water turbine driving twin ram pumps alongside Sleaford Navigation for supplying water to Haverholme Priory, Lincolnshire.

Hett's retirement may have been occasioned by a change of lifestyle after his marriage, but he also suffered from a weak heart. The business in which he was involved was highly competitive and he may have no longer wanted to use his energies in running such an operation.[52] Charles not only sold off the works, but, separately, he sold the good will, drawings and patents to one of his competitors, Gilbert Gilkes of Keswick, Cumbria.

The Ancholme works was not long out of use and after a short time it became home to the agricultural machinery dealer Peacock and Binnington, who still occupy the site.

Notes

[1] Charles Louis Hett, *Turbines of various types constructed at the Foundry*, (Brigg 1889), p.3; census return for Worksop, Nottinghamshire, 1861.

[2] Census return for Ripley, North Yorkshire, 1871; Lincolnshire Archive Office, YARB 68.

[3] LAO, DIXON 9/1/13/72

[4] 1861 census; *Lincoln Rutland and Stamford Mercury*, 22 April 1864, p.2, col. 5.

[5] *LRSM*, 4 March 1864, p.7, col. 5; *LRSM*, 10 July 1868, p.7, col. 2.

[6] E R Kelly, *Directory of Lincolnshire* 1876.

[7] Census returns for Brigg, 1871 and 1881.

[8] Census return for Brigg, 1891.

[9] *Kelly's Directories of Lincolnshire* for 1892 and 1896.

[10] *Minutes of the Proceedings of the Institute of Civil Engineers*, Vol 71, 1882-83, pp.163-175; Vol 80, 1884-85, pp.79-98; Vol 102, 1889-90, pp.183-189.

[11] C L Hett, *Rural water supply*, (London 1888); *A table of power of leather belting and shafts*, London 1889.

[12] C L Hett, *Engineering News*, 15 March, 5 April, 19 May and 17 May 1884.

[13] C L Hett, *Turbines*, 1887 and 1889.

[14] *North Lindsey Star*, 16 January 1892

[15] Marjorie J F Hett, *A Family History*, (Horncastle, 1934), p.80; *Kelly's Directory*, 1896.

[16] Patent number 13,035 of 1897 and 6981 of 1899, Patent Office London.

[17] Hett was one of 14 original members of the Zoological Photographic Club, which included RB Lodge, one of the pioneers of photographing natural history.

[18] See note 43 below.

[19] *LRSM*, 26 July 1872, p.3.

[20] *Kelly's Directories of Lincolnshire*, for 1876, 1885 and 1889; White *Directory of Lincolnshire*, 1882. Hett's cultivator is mentioned in an anecdote used in Edward Peacock's *A Glossary of Words used in the Wapentakes of Manley and Corringham, Lincolnshire*, (English Dialect Society, 1889). Tenant: 'That wood clooäse o' yours is strange unkind land. I oht to hev sum rent knock'd off o'count on it.' Squire: 'It isn't pulling rent off that will do it any good. It wants plenty of lime and Hett's cultivator through it twice over.'

[21] Lincolnshire Agricultural Society (LAS), *Subscribers List*, 1871.

[22] LAO, LRA 6/7/47, There was a total of six quotations and the total estimated cost of the bridge, including the brick piers was £700.

[23] Harry Symonds' Collection held at the Museum of Lincolnshire Life, Lincoln.

[24] F Henthorne, *History of Brigg Grammar School Trustees*, (Boston 1959), p.137.

[25] Edward Smith had been an ironmonger and iron merchant in Brigg since the end of the eighteenth century, see Pigott *Directory* 1794, 1822, 1835, 1841, and Kelly's *Directory* 1861.

[26] LAS, *Implement Catalogue*, Gainsborough, 1873.

[27] LAS, *Implement Catalogue*, Grantham, 1874.

[28] Henthorne, *History of Brigg*, p.98.

[29] C J Page, 'C L Hett' in *Lincolnshire Industrial Archaeology Journal*, 7, No 4, (1972). Day and Osgerby were manufacturing boiler composition material for removing and preventing incrustation, ie water softening.

[30] *The Implement and Machinery Review*, 3 August, 1876, p.595.

[31] *The Implement and Machinery Review*, 2 August 1878, p.1723.

[32] LAS, *Implement Catalogue*, Brigg 1880.

[33] *Implement and Machinery Review*, 1 June 1878, p.1610.

[34] *The Engineer*, 3 September 1880.

[35] Society for Lincolnshire History and Archaeology, survey by IA team

[36] LAS, *Supplement to the Implement Catalogue*, 1880; *Implement and Machinery Review*, 2 September 1880, p.3120; Wright, *Cyclopaedia*, 1911, Vol X, p.61 Fig.6.

[37] *Implement and Machinery Review*, Supplement, April 1879.

[38] *Implement and Machinery Review*, Supplement, April 1879; LAS, *Supplement to the Implement Catalogue*, 1880; C Page, *C L Hett*.

[39] *Implement and Machinery Review*, 2 September 1880, p.3119.

[38a] Wheeler, W H, *The Drainage of the Fens and Low Lands by Gravitation and Steam Power*, (E & F N Spon, London, 1888) p.102.

[40] Page, *C L Hett*; Reference in an unknown local county newspaper found in Lincolnshire Archives, 2 HETT 1/77 *Diary*, p.16.

[41] Wheeler, *The Drainage of the Fens*, p.153.

[42] ibid. pp.127-129; sale catalogue *Redbourne properties of Thomas Dann*, 1st September 1927.

[43] Gilbert Gilkes is a manufacturer of water turbines in Kendal, Cumbria.

[44] P N Wilson, 'Early Water Turbines in the United Kingdom', *Transactions of the Newcomen Society*, XXXI (1957 to 1959).

[45] LAS, *Implement Catalogue*, Brigg 1880; *Implement and Machinery Review*, 3 August 1880.

[46] Hett, *Turbines*, (1889), p.8.

[47] LAS, *Implement Catalogue*, Brigg 1889.

[48] Hett, *Turbines*, p.11; Wilson, 'Early Water Turbines in the United Kingdom'.

[49] Wilson, 'Early Water Turbines in the United Kingdom'.

[50] *Engineering*, August 1894; C Page, *'C L Hett'*; H D Gribbon, *The History of Water Power in Ulster*, (1969), p.244.

[51] *LRSM*, 12 August 1895, p.8, col. 5.

[52] LAO, 2 HETT 1/7/7. *Fanny Hett's Diary*, August 1911-January 1921, p2, 22 September 1911 'Dear Charlie died suddenly at Springfield. Heart troubles.....grief to us all, age 65'; 25 September 1911, 'Laid in our family grave in Brigg Cemetery'.

CHAPTER 9

WILLIAM HOWDEN & SON OF BOSTON
STEAM ENGINE PIONEER

Neil Wright

Lincolnshire deserves a place in the history of agricultural engineering as the county which made the first portable steam engine on wheels that could be moved from site to site without being dismantled. It was made by William Howden & Son of Boston in 1839. Other people had the same idea, but, so far as we know, Howdens were the first to actually make one. Also, so far as the history of engineering in Lincolnshire is concerned, Howden's portable engine was certainly the first steam engine to be made in the county.

William Howden (1774-1860) apparently opened his foundry near the Grand Sluice in Boston in 1803, though it was another 36 years before he made his major contribution to mechanical engineering.[1] Most of the great nineteenth century engineering firms of Lincolnshire were started in 1840 or later, and it has been suggested that Nathaniel Clayton (of Lincoln) and William Tuxford (of Boston) may both have got the ideas for their first portable steam engines from William Howden.[2]

For most of its early years Howden's business was a foundry producing iron and brass castings for all and sundry, rather than making steam engines or more advanced machinery. His products might have included domestic items such as stoves and grates, iron railings and gates for builders, agricultural implements and ironware needed by other trades in the town and port and surrounding area. Howden was responding to the local market and producing what was wanted, and when in due course steam engines were needed for local river boats, he was able to rise to the challenge. William Howden retired about 1855 (aged over 80) and his premises continued to be used for iron working until about 1880, the partnership including his son William until 1859.

Howden's Apprenticeship and Training

William Howden was born and bred in Scotland and we do not know exactly when or why he arrived in Lincolnshire, though what we do know suggests a likely explanation. He was born on 10 May 1774 at Cleish in Kinross-shire, about 20 miles north-west of Edinburgh, and by 1790 was apprenticed to Joseph Conachar, engineer and millwright, in the Scottish capital. Five or six years earlier, Sir Francis Kinloch, who lived near Edinburgh, had devised a threshing machine to thresh corn in barns, and one had been attached to a waterwheel. It was an improvement on a threshing machine made a little earlier by Andrew Meikle, a millwright near Dunbar, and Sir Francis had consulted Meikle about it. This was perhaps the first successful machine for the purpose, and Sir Francis wished to encourage other people to copy his idea.

In 1790 he placed an order with Joseph Conachar to make a working model of the threshing machine, at a scale of one inch to the foot. Mahogany was used for the woodwork and brass for the wheels and it was shown to be driven either by horse-power or a water-wheel. The job of making the model was given to young William Howden, working to instructions given by Sir Francis. So even as an apprentice he must have been showing great skill. Sir Francis was very pleased with the model, which he gave to the Agricultural Society at Bath. In fact he ordered two more from Howden; one was sent to America and the other to Russia.[3]

If Howden had started his apprenticeship when he was about thirteen and had continued for seven years, he would have finished about 1794. By 1802 he had arrived in the Boston area and married Ann Cicel at Wyberton, a village just outside the town. In between it seems that he worked for the great engineer John Rennie, who in the early 1800s was carrying out a number of jobs in the Boston area.[4] Rennie had also been born near Edinburgh, about thirteen years before Howden, and had been apprenticed to Andrew Meikle, whilst also attending classes at Edinburgh University, before moving to London. Rennie's great achievements in later years were in civil engineering, but his first major work was mechanical, the machinery for the Albion Mills at Blackfriars in London, built 1784-88 (but destroyed by fire 1791). While Howden was an apprentice he would have heard of Rennie's early achievements and perhaps saw him as a role model. He perhaps got a job with Rennie as soon as he was free of his apprenticeship, and so a little later came to Lincolnshire to assist Rennie there.

Soon after 1800 John Rennie was working on the project to drain the fens north of Boston and in 1804 he used the first steam engines in the fens, apparently to pump out excavations while cutting new drains. These engines were removed after the work was done.[5] Rennie also designed the new Town Bridge in Boston, which was not only the first cast-iron bridge he designed, but also the first cast-iron bridge in Lincolnshire.[6] Preparation work on the bridge started in 1802, which was the year when William Howden got married just outside Boston, and construction started in 1803, which is when Howden is said to have started his foundry.[7] Why did Howden set up in that business in that place at that date? Was it to cast elements of the new bridge; and did Howden then continue it as his own venture, seeing the potential for such a business in a flourishing and expanding town? These questions cannot be answered on the information we presently have. All that can be said with certainty is that Howden, apparently an employee of John Rennie, established one of Lincolnshire's first iron foundries.

The Foundry at the Grand Sluice

Howden's foundry was near the Grand Sluice in Boston on the east bank of the navigable river Witham *(Fig 114)*.[8] This area was to the north of the old town of Boston, and in the early 19th century it developed into a significant industrial suburb around the main landing place on the inland waterway.

In the 1760s the river Witham through the fens had been straightened and the Grand Sluice and the river above it were in a new channel created at that time. This section of river was parallel to Tattershall Road and the land between the river and the road was divided into plots that were developed for industrial uses and associated housing. Other businesses set up here over the following years included a boatyard, brewery, maltings, ropewalk, woad mill and the town's gas works, as well as one or two other small (and short lived) foundries. Since 1850 this part of Tattershall Road has been a cul-de-sac called Witham Town.

Pishey Thompson, the local historian of Boston, said in 1856 that Howden's works was called the Phoenix Foundry, but this name was not used in any directories referring to the firm.[9] After the works passed out of Howden's hands it was called the Boston Foundry in the 1856 Directory and the Grand Sluice Ironworks in 1861. In 1862 the site was temporarily split and the

Fig 114
Foundry House, now 11 Witham Bank East, (centre) from the other side of the Witham.
Howden's works was behind the house and behind the two tall trees to the right.
(Photograph 1964)

Fig 115
Counting House or office next to Foundry House.
(Photograph 2006)

other half was for a short time called the Witham Ironworks. The fact that it was the first foundry in Boston perhaps indicates that it did not need a name so long as Howden was running it, and that is probably confirmed by the use of the name 'Boston Foundry' in 1856.

The first written reference to William Howden's foundry site near the Grand Sluice was on 13 October 1817, when title deeds to the property to the south refer to this site as 'lands lately sold by Thomas Burrowes Burrowes to W. Howden'. Eight years earlier Mr Burrowes had purchased a square area of land that lay between Tattershall Road and the bank of the river Witham from a Mr Whiteman. By 1817, if not earlier, that large site had been divided into three strips, each extending from the road to the river. The land sold to Howden was the northernmost of the three strips.[10] It measured about 170 feet east to west by 70 feet north to south and is now the house and garden called 11 Witham Bank East. The 1809 deed did not say who the occupiers of the land were at that date, so we do not know for sure that Howden was there by that time. If the foundry started on that site in 1803,

then for the first fourteen years Howden was renting it from Mr Whiteman and then from Mr Burrowes. That does look possible, because in 1817 Howden was living in a rented house in Tattershall Road (now Witham Town) about 25 yards north of the foundry site.

Howden owned the rectangular site of his foundry from 1817 and he built himself a fine three-storey house, with a modest 'counting house' or office adjoining, at the end overlooking the river *(Fig 115)*. The counting house was at first floor level over a low flat-topped archway that gave access between the works and the river bank. The house and counting house still survive (2006) at 11 Witham Bank East and a date said to be carved on one of the roof beams of the house suggests it was built in 1820. The archway beneath the counting house was filled in after 1887. The 1826 *County Directory* lists William Howden as living on Witham Bank as well as having his business there. Censuses list him and his family living there in 1841 and 1851, and in the latter year it was actually called 'Foundry House'. He was probably living there from 1820 until about 1856. A plan of Boston in 1829 gives an indication of the layout of the foundry buildings behind his house at that date, and plans for proposed railways in the mid-1840s show that in the early part of 1845 the layout had not changed *(Fig 116)*. Shortly before 1839 Howden had purchased the parallel strip of land to the south of his site, though he seems not to have done anything to include it in his site for a few years.[11] Perhaps he just used it at first for the storage of raw materials. However, by the latter part of 1845 Howden had removed part of the eastern end of the dividing wall between the two sites and erected a building at the Witham Town end of the new land. So by the late 1830s the business was doing well enough to enlarge its site.

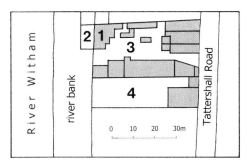

Fig 116
Plan of Howden's works in 1845.
1: Foundry House; 2: Garden in front of house, encroached on river bank; 3: Site of original foundry; 4: Extension bought c1839.

The Portable Steam Engine

County directories give some indication of the development of the business. In 1826 William was described as an engineer and iron and brass founder. By 1830 he had been joined by his son, William (born 1807), and the firm was also described as millwrights. After William junior got married for the second time in the 1840s, he moved out of the family home but still stayed in business with his father. The 1851 census listed the father as 'engineer and iron founder', and his son as 'millwright, engineer and iron founder', so evidently the son was looking after the millwrighting activities. That census indicated they were in partnership with each other and employed 21 men and three boys.

We have very little knowledge of the products this firm made because none of their business records survive. Probably Howden & Son's most important innovation was to produce the first steam engines to be made in Lincolnshire. Their first engines were small marine ones made in the 1820s for the market boats travelling along the river Witham between Boston and Lincoln. Howden's site on the bank of the river was ideal for this work. Clark suggests that Howden had probably started making the first such marine engines sometime in 1826 and the completed unit made a trial trip towards the end of 1827, whilst a longer voyage was made from Boston to Lincoln on 14 December of that year.[12] The boat was twenty-four feet long and the engine rated at two and a quarter horse-power. Engines for other river boats were made in succeeding years and in this connection Howden would have known Nathaniel Clayton, a young steam boat captain, who in 1842 joined with Joseph Shuttleworth to found the greatest Lincolnshire engineering firm of the nineteenth century.[13]

One of the great achievements of the nineteenth century was the invention of the traction engine, and a necessary precursor to that was the development of the portable steam engine, that is an engine on wheels that was light enough to be moved from place to place without dismantling. In the late 1830s, when William Howden senior was aged 65, the firm started to make portable steam engines that could be used in agriculture. He wrote a biographical note that was published on 28 September 1855 in the *Lincolnshire Chronicle* which said that:

> 'In 1839, I and my son made a Portable Engine of two horse power fixed on a frame with four carriage wheels, so that it was easily removed from one place to another by horses, and employed in thrashing and other purposes; this was the first of the kind, and we claim the invention. In 1841, we made one of six horse power, which was shown at the Agricultural Meeting at Wrangle, and for which we were awarded the prize of two pounds; since which time a great many have been manufactured both for home and foreign orders.'

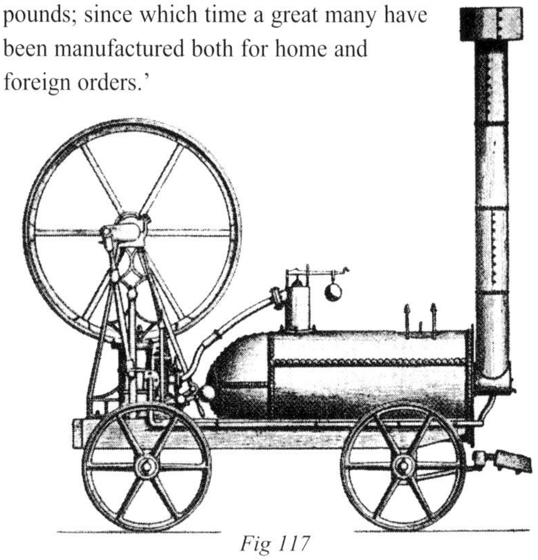

Fig 117
Howden's portable engine of 1839 as illustrated in Fielden's Magazine.

A few portable engines may have been made by various small country engineers in the 1820s, but Howden's is the first which we know was definitely built and sold, so perhaps William Howden does deserve the prize as builder of the first portable steam engine *(Fig 117)*. It appears that he did not fully appreciate the importance of his own invention, because it is said that he only built twelve portables altogether and then gave up their manufacture because he was afraid the country would become overstocked![14] As already noted, William Tuxford and Nathaniel Clayton may well have obtained the ideas for their own first portables from William Howden's 1839 model, Tuxford making his in 1842 (from designs of 1839) and Clayton's appearing in 1845.[15]

Other Products of the Howden Foundry

The firm also claimed to have made the first portable threshing machine. Pishey Thompson wrote in 1856: 'We believe the first moveable steam threshing machine ever made was constructed at the Phoenix Foundry in 1841.' This claim was not included in William Howden's biographical note, but was included in their entry in the 1849 *Directory* where they described themselves, among other things, as 'inventers and makers of the portable steam threshing machines'. The other important Boston engineering firm was not far behind Howdens. Pishey Thompson said that soon after Howden's invention Tuxfords of Boston made the first moveable combined threshing and dressing machine by steam power.[16] *(see p134)*

Other evidence shows that Howden & Son made a wide range of products, including the cast-iron pillars for Centenary Methodist Church in Boston, built in 1839-40. In 1844 they installed a cast-iron bell frame for the church bells at Wainfleet St Mary and another bell-frame at Wigtoft in 1858. William senior might also have been responsible for the partial cast-iron frame installed at Wrangle church in 1822, which John Ketteringham suggests was perhaps the earliest use of iron for a bell frame in the country.[17] They also made cast-iron ship carriages for three seventeenth century cannons belonging to Boston Corporation *(Fig 118)*. The date cast into the carriages looks like 1848, but may be 1843, which was the 200th anniversary of the acquisition of the cannons.

*Fig 119
Turnpike Mile Post made by Howden, restored by Rundles of New Bolingbroke in 2006 and reinstated by Lincolnshire County Council near its original site on the A17 near Algarkirk.*

*Fig 118
One of the carriages made for Boston Corporation by Howden in 1843 or 1848 to carry Civil War cannons.*

Howden also made cast-iron mile posts for the Swineshead to Fosdyke turnpike trust (formed 1826), of which at least two posts survive, one still in a cut-off section of road at Drayton, Swineshead, and another *(Fig 119)* restored by Rundles of New Bolingbroke in 2006 and reinstated by Lincolnshire County Council near its original site in Algarkirk.

Decorative ironwork still remains around the front of Foundry House and its neighbour (11 and 13 Witham Bank East). In addition to these specialised jobs, Howdens would also be producing agricultural implements and domestic items such as kitchen ranges.

Howdens seem to have been at their most productive from the late 1830s to the early 1850s when both father and son were involved. Their works were extended at this period and their entry in the 1849 *Post Office Directory for Lincolnshire* is the most detailed: 'Howden & Son, iron & brass founders, engineers & millwrights, steam engine boiler manufacturers, & inventers & makers of the portable steam thrashing machines, near the Grand Sluice'. But the firm seems to have lost dynamism about that time, perhaps due to the age of William senior, and although the Great Northern Railway passed close to the works, the firm did not get a railway siding into its premises. From January 1848 there was a contractors' railway line along the bank between Howdens' works and the river, necessary for the building of the GNR and East Lincolnshire lines, and from October 1848 to May 1850 that 'temporary' track was used for passenger and goods trains until a new embankment could be built on the other side of Tattershall Road.[18] That must have made it very difficult for Howden to receive raw materials by river for those two and a half years.

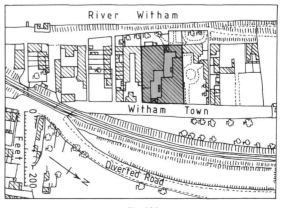

*Fig 120
Part of 1887 Ordnance Survey Plan of Boston showing the Grand Sluice Ironworks (disused at that date).*

The 1887 Ordnance Survey map showed that since 1845 there had been considerable development on the site of Howdens' works, with buildings covering not only most of the original site but also most of the additional site to the south *(Fig 120)*. It is not clear whether this expansion occurred in the late 1840s/early 1850s under Howden ownership, or was carried out ten years or so later under William Wilkinson or Henry Wright.

William Howden: the Man and His Family

There is little, if any, evidence that either William Howden or his son took any part in Boston's affairs outside their foundry. William senior had married in 1802 and his only son seems to have been William junior who was born in 1807. In the 1850s to 1890s, there was a Robert Howden in Boston, who in the 1850s to 1870s was an engine fitter, but he had been born in Sheffield in 1818 and no evidence has been uncovered to connect him with Howden & Son. William junior had married by 1838 and had a son Robert born in that year, but his wife died, aged 22, on 23 May 1839. Before 1851 he had married again, his second wife Sarah having been born in Stepney, London, in 1823. In 1851 William and Sarah were living in Spain Lane, Boston, but 13-year-old Robert was living with his grandparents on Witham Bank on census night. No record has been seen of any children from William junior's later marriage, so both Howden households were small by Victorian standards.

This was perhaps why William senior had few servants, even though he had a big house and was head of a medium-sized business. In 1841, when all three generations were living in Foundry House, they only had two servants living-in. Ten years later in 1851 William junior had one resident female servant in his small house in Spain Lane and his parents had reduced their establishment to one resident servant as well.

We know little about William Howden's church attendance or other activities. As William senior had not been born in Boston, nor served his apprenticeship there, he was not entitled to vote in local or parliamentary elections until 1832. Poll books show that when he and his son both got the vote they supported the Whig, Radical or Liberal candidates. There is no evidence that they got actively involved in local politics, unlike, for example, the Tuxford family, who ran Boston's main Victorian ironworks, or William Wilkinson, who later took over Howdens' works. In fact William Howden senior did not even vote in 1832, but only used his parliamentary vote for the first time in the 1835 general election. This gives the impression of a frugal Scottish family, keeping themselves to themselves.

By 1856 the founder of the firm was over 80 and had evidently retired, though he continued to live in Foundry House. William Howden senior died on 7 May 1860 aged 86 and the cause of death was said to be 'senile decay'. By that date the family was no longer involved in the foundry and William died in a house in Red Lion Street, closer to the town centre.

The Last Days of the Business

In the 1856 *Directory* William Howden junior of the 'Boston Foundry' was listed as an iron and brass founder, machine maker and millwright, but he did not last very long on his own. From 25 February 1856 he entered into a partnership with William Wilkinson, John Short and Henry Thompson Wright 'in the business of ironfounders and engineers for five years', and in September 1856 William Howden senior conveyed the property to the new partnership.[19] Wilkinson was apparently the captain of a steamship and a trader on the River Witham and was the dominant partner in the new firm. For a time they traded as Howden, Wilkinson & Co. until Howden left on 28 February 1859, conveying his share of the business and the property to Wright and Wilkinson, who then traded as Wilkinson, Wright & Co.[20] We do not yet know what happened to young Howden after he left this business; at the time he would be aged 52, so could have continued working for some time.

The new partnership apparently brought fresh vigour to the Boston Foundry, or Grand Sluice Ironworks as it became known. The 1861 *Directory* described the firm as: 'Wilkinson, Wright & Co., engineers, millwrights, iron and brass founders, and manufacturers of portable and fixed steam engines, thrashing machines, straw elevators, brick & tile machines, saw tables, & hydraulic machinery of every description, Grand Sluice Ironworks'. The census listed Wilkinson as 'engineer employing 50 men and 20 boys', which, if true, was twice as many employees as Howden had ten years earlier. But this partnership was also short-lived and ended on 28 February 1861. They divided the premises between them and each carried on trading as a separate enterprise, Wilkinson as 'engineer, millwright and iron and brass founder' in the northern half, called the Witham Iron Works, and Henry Thompson Wright as 'ironfounder and engineer' in the southern half, still called the Grand Sluice Iron Works. The new arrangement lasted less than three years, and in December 1863 Wilkinson moved to the Poole Foundry in Dorset with Stephen Lewin of Boston as his sleeping partner.[21]

Wright took over the whole Howden site again and in 1872 his directory entry was: 'Wright, Henry Thompson, ironmonger [C & T & H T Wright] and ironfounder, engineer, millwright, &c., Grand Sluice and Witham Town ironworks'. It is said that Henry Wright was well known as a maker of agricultural implements *(Fig 122)*.[22] The previous Boston partnerships had been financial failures and it was left to Wright, with the help of two members of his family, to gather together the assets of Howden, Wilkinson & Co. and Wilkinson, Wright & Co. in an attempt to discharge the partnership debts. In the end there was a shortfall in the amount due to creditors, which the Wright family made good from their own monies. They later attempted to recover from the former partners their share of the debts, but in this they were not particularly successful.[23]

Henry Wright continued to operate the Grand Sluice Ironworks until shortly after 1876 and the works was not listed in any subsequent directories. Foundry House became known as 11 Witham Bank East and the Wright family owned the site until September 1904.[24] The works was shown as disused premises on the 1887 OS plan and shortly after that the buildings to the rear of the house were demolished and have stayed as a garden to the present day. The southern section, at one time the Grand Sluice Ironworks, was sold off separately and has continued in light industrial use until recent times with the address of Witham Town *(Fig 121)*. Howdens' business, continued by Henry Wright for about 20 years, came to an end over a century ago and there are now few signs that this small site once played a key role in the industrial revolution in Lincolnshire and produced the first steam engines to be made in the county.

Fig 121
East frontage of Howden's site in Witham Town - the original site to the right (behind the car) is now a garden and the extension to the left has some buildings that may date to Howden's time. (Photograph 2006)

Notes

[1] P Thompson, *History and Antiquities of Boston*, (1856, reprinted 1997), p.348.

[2] Ronald H Clark, *Steam Engine Builders of Lincolnshire*, (1955, reprinted 1998), pp.60, 61.

[3] *Lincolnshire Chronicle*, 28 September 1855, p.6.

[4] P Thompson, *History and Antiquities of Boston*, (1856, reprinted 1997), p.348.

[5] Dr R L Hills, *The Drainage of the Fens* (2003), p.70.

[6] B Barton, 'Bridging the Gap' in *Lincolnshire on the Move*, (2005), p.54.

[7] P Thompson, *History and Antiquities of Boston*, (1856, reprinted 1997), p.348.

[8] *Ibid.*

[9] *Ibid.*

[10] Title deeds to 11 Witham Bank East, Boston.

[11] *Ibid.*

[12] Ronald H Clark, *Steam Engine Builders of Lincolnshire* (1955, reprinted 1998), p.60.

[13] Neil R Wright, 'William Howden & Son, Engineers of Boston' in 'Industrial Archaeology Notes' edited by Mark Bennet in *Lincolnshire History and Archaeology*, Vol.33, (1998), p.59.

[14] W J Hughes, *A Century of Traction Engines,* (1959), pp.7, 8.

[15] Ronald H Clark, *Steam Engine Builders of Lincolnshire*, (1955, reprinted 1998), pp.60, 61.

[16] P Thompson, *History and Antiquities of Boston*, (1856 reprinted 1997), p.348; *Post Office Directory of Lincolnshire*, (1st edition, 1849), p.2,988.

[17] John R Ketteringham, *Lincolnshire Bells and Bellfounders*, (2000), p.263.

[18] Neil R Wright, *The Railways of Boston*, (1971, 2nd edition, 1998), pp.20, 25.

[19] Russell Wear and Eric Lees, *Stephen Lewin and the Poole Foundry*, (1978), p.13.

[20] *Ibid.*

[21] Op. cit., p.19.

[22] Ronald H Clark, *Steam Engine Builders of Lincolnshire*, (1955, reprinted 1998), p.132.

[23] Russell Wear and Eric Lees, *Stephen Lewin and the Poole Foundry*, (1978), p.13.

[24] *Ibid.*

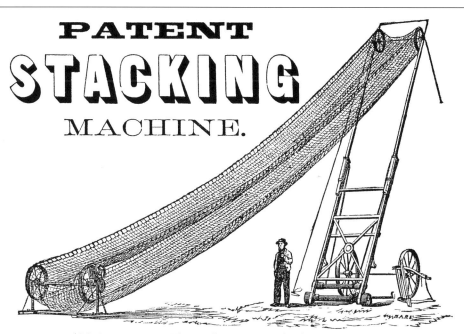

Fig 122
Henry Wright occupied the Grand Sluice Ironworks after Howden & Son

CHAPTER 10

THE MALLEABLE OF NORTH HYKEHAM

CASTINGS FOR AGRICULTURE AND THE MOTOR INDUSTRY

Norman Tate

The Malleable had its origins in 1874, when three local businessmen Frederick Harrison, John Teague and James Birch decided there was a call for a foundry to supply castings to the burgeoning agricultural engineering industries that had sprung up in Lincoln. At this time Ruston, Proctor & Co, Robey, Clayton and Shuttleworth, and Foster were already well established, but, of the three, Ruston, Proctor & Co was the only one to have its own foundry. As a consequence there was a demand for good quality malleable iron castings for local companies and also for the thriving and expanding railway industry.

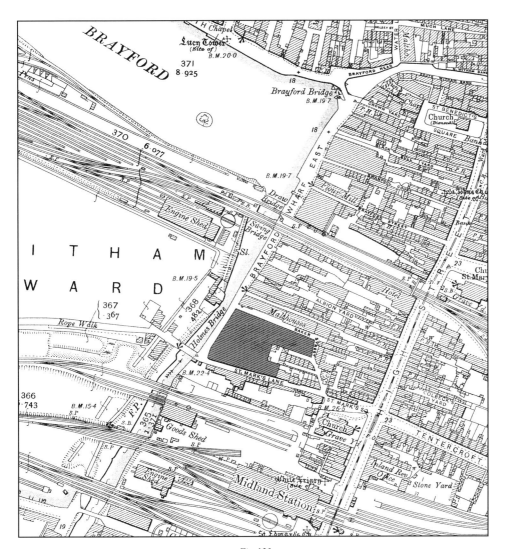

Fig 123
Site of the Malleable Iron Works in Lincoln (1874-1922) from OS Plan of 1905.

Harrison & Co (Lincoln) Ltd

In order to establish the new foundry, Harrison persuaded Teague and Birch to join him in purchasing the site of a former millwright in the St Marks area of Lincoln. This was on the corner of St Marks Street and Brayford Wharf East *(Fig 123)* and was until recently the site of the Lincoln Corporation bus depot. The site is occupied in the early twenty-first century by a large block of student accommodation known as 'The Junction'

The new company began producing 'whiteheart' malleable castings, very quickly became successful and was registered as Harrison & Co (Lincoln) Ltd in December 1904. In 1905 it became obvious that more production space was needed by the growing company, but none was available on the Lincoln site. Therefore a parcel of land, a former jam factory site, was acquired in North Hykeham next to the railway station on the Midland Line and a small branch works established. These were the works that were to become known to the local people as 'The Malleable' *(Fig 125)*.

The success of the firm continued and the Hykeham branch grew. Castings for gear wheels, spokes, axles and other agricultural machine parts were being made at this time in large quantities for local firms such as Robey and Penney and Porter. The company also began to supply castings to the expanding Midlands motor trade, which included household names such as Austin, Morris, Sunbeam, Triumph, Vauxhall, Bedford and Leyland among its customers.

Fig 124
Alice Bradford, a worker in the moulding shop, c1915.

Fig 125
Foundry buildings and workers at North Hykeham shortly after the new site opened in 1905.

Fig 126
Workers at North Hykeham with agricultural machinery castings, 1910.

In 1922 the Lincoln site was closed and all operations were transferred to North Hykeham. A period of rapid growth and investment followed.

In the early days production was limited to 'whiteheart' malleable iron, which has only marginally more elasticity than cast iron, but from about 1930 the motor trade was demanding castings with a greater level of ductile strength.

The meaning of the term 'malleable iron' needs explanation. Ordinary (or white) cast iron, which was one of the essential ingredients of the industrial revolution, is very brittle. It can break if it is dropped or struck by another hard object because it has very little elasticity or ductile strength. Malleable iron is created by re-heating white cast iron to a very high temperature over an extended period of time and then allowing it to cool gradually. This process has the effect of giving some elasticity to cast iron. The degree of elasticity can be varied according to the mixture of constituents that are put into original production of cast iron and by different types of re-heating afterwards.

The advent in 1930 of rotary oil-fired furnaces enabled Harrisons to switch from production of only 'whiteheart' castings to what became known as 'blackheart' castings, which has greater ductile strength. This type of casting was in great demand for the motor trade. Harrison & Co continued to expand until the next stage in its development in 1938.

Leys Malleable Castings Ltd

In 1937 the long-established and much larger firm of Leys Malleable Castings Ltd of Derby bought Harrison & Co. Co-incidentally, Leys had also been formed in 1874, when Francis Ley, the founder, left his father's farming business to make ductile castings for the agricultural industry.

Early efforts were not successful until he went to America to learn the secret of 'blackheart' malleable production, which had been discovered in 1820 by an American, Seth Boyden, who had established the American Malleable Iron Foundry at Naugatuck, Connecticut, USA. Francis Ley obtained a licence to use this system in the UK and this set him on the road to becoming the largest producer of malleable castings in this country.

In some ways the rise of Leys in Derby mirrored that of Harrisons in Lincoln in that Francis Ley set up his foundry in the city of Derby and then ran out of room for expansion. Unlike Harrison, instead of finding new premises a short distance away on the outskirts of the city, he looked around for an established company elsewhere to take over.

Fig 127
Foundry workers at North Hykeham taking a break, 1910.

Harrisons' works, situated next to the main railway line between Lincoln and Derby, proved to be ideal. That there was also a good road link (for those days) between the two works was an added bonus.

Significant numbers of senior management personnel were re-located from Derby to Lincoln and, by the time of the outbreak of the Second World War in 1939, massive investment was beginning to take place.

Fig 128
Lincoln Malleable Iron Works Band, founded 1893. F H Harrison, the firm's founder, is fourth from the right in the front row.

Fig 129
Tapping a furnace into a ladle, 1958.

New, modern, semi-automated machine moulding lines were installed in 1939, 1941 and 1944 to replace the very labour-intensive hand floor-moulding used previously. Also at that time the means of melting pig-iron, which forms the base for malleable iron, significantly improved. The smaller furnaces producing nine tons per hour were coal-fired, stoked by hand using lump coal, but the larger furnaces producing twenty-three tons per hour were using pulverized coal dust fed by air-pressure. This resulted in much higher temperatures, thus producing molten iron more quickly and enabling the production of molten metal to keep up with the speedier production of the semi-automated moulding lines.

As can be appreciated, use of coal dust and black moulding sand made the foundry a very dirty working environment. Leys had had a good reputation in Derby for looking after its employees and for being generous philanthropists, and in North Hykeham one of the first improvements introduced was an on-site shower block with locker rooms, so that workers could shower and change and go home clean. The company also encouraged a wide variety of social activities for its employees and their families.

Returning to the changes taking place in production techniques, 1949 saw another significant improvement with the installation of gas-fired annealing of the castings. This meant the castings could be stacked up on a firebrick base and a gas-fired oven lowered over them by overhead crane and then connected to the gas main and fired-up. This replaced the previous unwieldy process of packing the castings into small pans with iron ore and feeding these into pulverized-coal ovens. The time of the annealing process was also reduced from eight days to three days.

1954 saw the installation of the first fully automated moulding line, which proved to be successful after some initial teething problems, but in spite of this a still significant amount of floor and bench moulding was taking place. At this time The Malleable had a total of about 850 employees.

Fig 130
Hand casting on the B-line using ladle on the mono-rail, 1964.

Fig 131
Chipping off or fettling (cleaning) the casting for a lorry axle, 1964.

1962 to 1964 saw the undertaking of a complete modernization programme. The automated moulding line, which made large castings, installed in 1954, was designated 'A Line'. Two more automated lines, designated 'B Line' and 'C Line', each making smaller castings, were added. Finally, 'D Line', making really large castings, such as differential casings for Bedford lorries, was introduced.

In order to cope with this huge increase in production, all the coal-fired furnaces were replaced by one large central Duplex hot-blast furnace capable of producing fifteen tons per hour for sixteen hours per day. At this point the foundry production staff moved to two-shift working (6.00 am to 2.00 pm and 2.00 pm to 10.00 pm) and the in-house maintenance workers moved to a three-shift system, with a permanent night shift to do routine maintenance when the foundry and melting plants were shut down.

In 1966 the automated 'A Line', which had started in 1954, was closed and completely rebuilt as a self-contained, fully-automatic foundry with its own melting plant of three electric induction furnaces supplied by Brown Boveri of Switzerland. Also at this time a brand new continuous gas-fired annealing furnace capable of producing 200 tons per week of pearlitic malleable was built. As demand from the motor industry both here and abroad continued to increase, a further two cold-blast cupolas were added to the electric induction furnaces.

By this time, in spite of all the automation, the workforce had grown to 1,100 personnel and it needed both the Lincolnshire Road Car Company and British Rail to get the workers to North Hykeham from Lincoln and the surrounding villages.

Unfortunately, this upward spiral of investment and increased production could not last. Throughout the 1970s, as the British motor industry began to decline and collapse, demand for malleable iron products reduced accordingly. This resulted in the workforce being reduced by redundancies and money being unavailable for continued investment and upgrading of plant.

Fig 132
Completed castings being loaded on the company's lorry for despatch, 1964.

Fig 133
Aerial view of North Hykeham works from west, c1997.

Up to this point, neither Harrisons nor Leys had found it necessary to employ sales people, because the motor industry had always beaten a path to their doors; demands for their products had always been so high that a sales force had been unnecessary. With the steady globalisation of the motor industry, motor manufacturers could now source castings abroad considerably cheaper than Leys could produce them. As production at North Hykeham diminished, it looked as though Leys would focus their efforts on their Derby headquarters and the future for the North Hykeham plant looked bleak.

In 1979 more job losses were announced, but salvation arrived from an unlikely source. On 1 October 1979 Leys Malleable Castings Ltd announced that it had reached agreement with the Georg Fischer Group of Schaffhausen in Switzerland to form a joint venture for producing a new type of casting called spheroidal graphite iron (known as SG iron). The announcement was made that Georg Fischer had taken a 50% shareholding in the North Hykeham works, which became known as Leys George Fischer (Lincoln) Ltd, but this arrangement lasted for only one year.

George Fischer (Lincoln) Ltd

In October 1980 Georg Fischer acquired Leys' remaining 50% share of the business and the firm became a wholly owned subsidiary of the Georg Fischer Group under the name George Fischer (Lincoln) Ltd. One of the credos of the Georg Fischer Group was that the time to invest and modernize was during recession and downturn in business. As the acquisition of the North Hykeham plant occurred at the time of a very severe recession in the foundry industry, Georg Fischer set about investing huge sums of money in the North Hykeham plant in order to be ready when the expected upturn came.

In the first two years a total of £30 million was invested in new plant and in upgrading the working conditions for the now depleted work force. Incidentally, after the sale of the North Hykeham works to the Georg Fischer Group, Leys retreated to their Derby headquarters and attempted to consolidate. Unfortunately this lasted only until 1984 when the Derby foundry was sold and by 1986 it was totally closed down.

Fig 134
Charging the GED furnace for new moulding line with the aid of Kalmar truck, c1997.

Fortunately for Lincoln, the North Hykeham foundry survived the recession and, thanks to continued investment from Georg Fischer, began to thrive again. In 1982 a brand-new multipurpose moulding line was installed with fully-automated metal pouring. At the same time the large melting plant, installed in 1963, was fully modernized to support the new moulding line. Also in 1982 the three remaining moulding lines and the second 1969 melting plant were closed down. About this time, too, a contract was obtained from the Ford Motor Co. to produce and supply aluminium castings for the Sierra range of cars.

The production of aluminium lasted only until 1989, but other investment continued apace. The outside of the foundry was completely encased with metal cladding to give a more modern appearance. This was after significant investment to reduce emissions of various particles from both the melting plants and the casting processing plants to comply with EEC regulations. The emissions from the foundry had been a bone of contention between the foundry management, local residents and the local councils since Harrison had first opened the foundry in 1922!

The demands of the buyers of castings for much better finished products, rather than the rough finish that had always been acceptable in the past, led to investment in more sophisticated processing plant so that the buyer had to do less with the casting when he received it. This led to the installation of sophisticated shotblasting plant, painting facilities and even computer-controlled machinery.

From 1996 to 1997 an investment of £30m in a new melting plant, moulding line and trimming line meant an increase in capacity to 77,000 tonnes of good finished castings. By 2000 the company employed about 600 people and was exporting 90% of its cast products for the production of Volvo, Daf, and Daimler/Chrysler vehicles. Later increases in automation led to decreases in staffing levels to 400 people. (The foundry had once employed over 1,000 staff.) This level of staffing continued until 1 May 2004 when it was announced that George Fischer (Lincoln) Ltd was being sold.

Lincoln Castings: the Final Two Years

The sell-off was made to the Meade Corporation, a privately-owned engineering group, which renamed the North Hykeham foundry Lincoln Castings. The larger moulding line installed in 1996 was removed and transported to Germany and the workforce reduced to approximately 200. For a little over two years this company continued to supply high quality castings to the truck industry, but also found new markets in the rail and diesel engine industries, producing cast products such as cylinder heads for the first time in its history. Production finally ceased on the Hykeham site in December 2006.

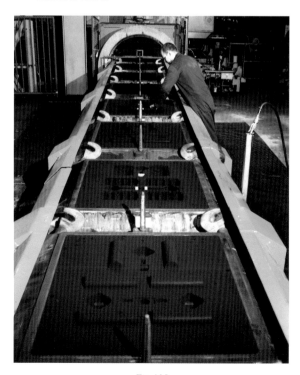

Fig 135
LM2 moulding line, c1997.

Postscript: Social Activities

Early in the twentieth century Harrisons purchased land to the rear of the foundry and turned it into a sports field. This site, which was approached by a track off Grace Road (Station Road), continued as a sports amenity until it was sold off for sand and gravel extraction after the Second World War. During the pre-War period, The Malleable produced many successful football, cricket, tennis and bowls teams which competed in the local leagues. In the 1960s Leys bought a large field opposite the foundry on Station Road and developed this as a new sports facility.

Harrisons, in their day, also supported another social activity, the works brass band. The Lincoln Malleable Iron Works Brass Band, founded in 1893 *(Fig 128)*, enjoyed much success, culminating in 1920 when the band reached the final of the All England Brass Band Championship at the Crystal Palace and finished in second place. Regrettably, in 1936 the firm dropped its sponsorship, but the musicians reformed themselves as the Lincoln Borough Silver Band.

Fig 136
Lorry wheel hub and lorry brake carrier, cast at North Hykeham, c1997.

Fig 137
Melting plant, showing molten iron pouring from the cupola into the holding furnace, 1997.

CHAPTER 11

PEACOCK & BINNINGTON OF BRIGG

AGENTS FOR TRACTORS & FARM MACHINERY

Philip Brown

Henry Peacock and the Early Days in Hull

1894 was a good year for openings, especially those of iconic feats of engineering. In May Queen Victoria opened the Manchester Ship Canal; June witnessed The Prince of Wales opening the Tower Bridge over the Thames; and in September Blackpool Tower, the seaside resort's replica of the Eiffel Tower, was opened, which, at 518 feet (158 m), was then Britain's tallest structure.

In the same year a much smaller engineering opening happened in modest premises at 11-15 West Street, Hull. The partnership between Henry Peacock and John Binnington may not have grabbed national headlines, but their newly founded firm, Peacock and Binnington, began and later progressed to become a major influence on the Lincolnshire agricultural industry for over one hundred years and still remains a force in the market place today.

Henry Earl Cartwright Peacock was born at Hay-a-Park, Knaresborough, on 10 March 1866. He came from a good tradition of Yorkshire farmers including his great-uncle, Henry Tennant Peacock, who was both an arable farmer and a breeder of pedigree shorthorn cattle near Ripon. Henry Earl Cartwright's father, Dennis Peacock, was both a farmer and an agricultural journalist until poor harvests, caused by exceptionally adverse weather conditions, contributed to his going out of farming in 1879. In 1881, after Henry had left Knaresborough Grammar School at the age of fifteen, Dennis decided to apprentice him to Lincolnshire ironmonger and close friend of the Peacock family, Joseph Cartwright of Long Sutton. A close relationship is suggested by the incorporation of Cartwright in Henry's name. In July 1885 Joseph Cartwright wrote to Dennis Peacock stating that, 'I have pleasure in expressing my entire satisfaction of Henry's general conduct and the manner in which he has discharged his duties during the said term and shall always hope for his advancement and prosperity.' This endorsement of Henry's skills and work ethic could be said to have marked the transition of the Peacock family from farming the land to agricultural engineering.

After leaving Long Sutton Henry continued in the ironmongery trade for the Leeds based firm James Nelson and Sons, before, at the age of twenty-three, becoming a junior partner in a short-lived firm named Hirst, Leech (also a former James Nelson employee) and Peacock. They opened premises in 1889 and mainly sold safes and locks as agents for Chubb and Sons. After a stay of three years Henry Peacock left Leeds to take up a managing role in Hull for the Hull Cart and Wagon Company, constructors of carts, wagons and milk floats. Henry's position in the company required his attendance at Hull Docks, often as early as five am. It was here that fate dealt its hand as Henry met John Binnington, a man of similar views and work ethic, and in 1894 the partnership of Peacock and Binnington, agricultural engineers, was created. A 1980 Peacock and Binnington newspaper described the two men as: 'one a tall, rather studious, bespectacled man (Binnington), the other a stocky, solid example of Yorkshire grit (Peacock).'

Fig 138
Henry Earl Cartwright Peacock (1866-1938).

The business began at 11-15 West Street, Hull. Cook's *Hull Alphabetical Directory* for 1897 lists 'Peacock and Binnington, agricultural implement merchants', as well as 'Peacock, H E C [P & Binnington] 16 Albert Terrace, Anlaby Road'.[1] They later moved to 14-16 Pease Street, Hull, which was situated next to Albert Terrace. (The first trade directory listing at this new address is 1913.[2])

An insurance document drawn up by The Ocean Accident and Guarantee Corporation Limited, dated September 1919, described the property insured: 'On the Buildings, communicating, brick corrugated iron and partly timber built and roofed with slates, corrugated iron, timber and felt, situate Nos.14 and 16, Pease Street, Anlaby Road, Hull, aforesaid, occupied by the Assured as Office and Warehouse for Agricultural Machinery and Implements (containing a pipe stove for warmth and two woodworking benches).' The annual premium was £3.7.6 and the premises were insured for £1,500. A proviso was that the stove should be kept clear from all woodwork and other combustible material and that all shavings and other refuse be swept up and removed at least twice a week. Their main line of business at this time was the sale of ploughs, threshing machines and horse drawn carts. The small size of the premises, with only two work benches mentioned, would suggest that the bulk of the sales were other manufacturers' machinery, with only smaller items such as sack lifters and barrows that they themselves made.

Henry Peacock Establishes the Business in Brigg

The first year of trading must have been very difficult; the 1893/4 winter was so severe that the River Trent was frozen solid enough to allow horse-drawn wagons to cross the ice. However, with Henry Peacock's experience and foresight bolstered by John Binnington's financial backing, the fledgling firm established itself, and only two years later the search began for a suitable place to expand the business. The popular north Lincolnshire market town of Brigg (still boasting five agricultural engineers today) had always been a magnet and meeting place for the area's farming community. Witnessing the packed and busy activities of a Thursday market day, Henry Peacock identified the town as a perfect location for a new depot south of the Humber. A local inventor and engineer, noted for his manufacture of vertical steam engines and pumps, Charles Louis Hett, had recently retired and his small foundry rented from Lord Yarborough had become vacant.

So it was here, on The Old Foundry Site in 1896, that Henry Peacock initially set up an agricultural machinery spares department, followed soon later by machinery sales, in the area where Peacock and Binnington still stands today. A plaque acknowledging the change of business from Hett to Peacock and Binnington can be found in their main Brigg showroom *(Fig 139)*. *Kelly's Directory of Hull & Lincolnshire*, 1900 lists: 'Peacock and Binnington, agricultural implement manufacturers & cart & wagon builders, Foundry Lane [Brigg]; & 11 to 15 West Street, Hull'.[3] Unfortunately 1896 also suffered from adverse weather conditions. An article under the headline 'Gloomy Agricultural Outlook in Lincolnshire' in the *Lindsey and Lincolnshire Times* dated Saturday, 19 September 1896 stated that: 'It is now more than three weeks since any harvest work was practicable, and during that time heavy rain has fallen almost every day. To many farmers this will be the most disastrous harvest experienced for many years'.[4] But with hard work and great dedication the business managed to keep afloat.

Fig 139
*Pulley from Hett's Ancholme Foundry,
the site now occupied by Peacock & Binnington.*

From the beginning Henry Peacock's business philosophy was squarely based on good customer relations and quality of service. Not only did he travel many miles throughout Lincolnshire, but he also made sure that the business was well represented at local agricultural events such as the Lincolnshire Agricultural Show and the Yorkshire Show. To make himself more visible at the markets and shows, Henry, a tall man, chose to wear distinctive head gear. 'Henry clearly believed that it paid to advertise and made himself conspicuous as "the man in the half-crown hat", that he used to wear to markets to stand out in the crowd.'[5]

The firm's first appearance at the Lincolnshire County Show was in 1898, where their stand displayed binders by Massey-Harris, Deering and Plano, a Ransome cultivator and two digging ploughs. In later years they also exhibited 'made by exhibitor' products such as spring pulleys, sack lifters and sack barrows.

An extract from the *Dewsbury News and Chronicle* in 1900 reported on the Royal Agricultural Show at York and stated that: 'Messrs Peacock and Binnington of 11 to 15 West Street Hull, in the implement sheds at Stand No.130, had an excellent and useful collection of self-binders, reapers, mowers, horse rakes, carts, rullies and milk floats.'[6] Their other exhibits included cream separators, churns, sack barrows and other farming requisites. They also felt it very important to get themselves known and recognised amongst the wider implement manufacturing communities at national shows such as Smithfield and the Royal Show.

Peacock and Binnington also made sure that they had a dominant presence in Brigg every Thursday on market day. A wooden hut was towed the short distance over County Bridge across the river Ancholme and positioned in the Market Place. This acted as both a display facility and a meeting place between representatives of the company and the farming community and was where sales and deals frequently took place.

Fig 140
Peacock & Binnington's former showroom and retail premises on the corner of Bridge Street and Foundry Lane, Brigg.

Fig 141
Peacock & Binnington stand at the Yorkshire Show, 1905.

A photograph dated 1914 shows this hut advertising not only Ransomes, Sims and Jefferies' goods, but also lists them as agents for Bamford's mowers and reapers. This practice continued for many years and a painting of Brigg Market Place of the 1960s, hanging in The Angel function room entrance, clearly shows Peacock and Binnington's red hut.

By visiting farms and talking to customers, Henry Peacock's engineering background helped him to recognise the needs of local farmers. After obtaining the area's sole dealership rights for the prestigious implement manufacturer, Ransomes, Sims and Jefferies of Ipswich (mainly due to the badgering of Mr. Ransome at the Smithfield Show), Henry persuaded them to make modifications to their implements to meet local farming conditions. So great was his input into some products that Henry Peacock was able to earn royalties on their sales. An example of a modified Ransome plough with a Peacock and Binnington stamp still stands proudly at the front of their Brigg depot today. An advert in the 1907 *Jackson's Brigg Annual* identifies Peacock and Binnington as agents for Ransomes' Chilled Ploughs and Patent Chilled Plough Shares, asserting that: 'local-made ploughs are the very best.'[7] The advert also carries an unnamed picture showing that the company was also now involved in selling steam-driven traction engines of unknown manufacture. A sales leaflet, possibly from the 1930s and giving Peacock and Binnington's telephone number as '13 Brigg', advertises Ransomes' ploughs of varying sizes, costing from £4 up to £13 *(Fig 142)*. In some cases it lists the number of horses required to pull them efficiently, horsepower meaning literally that in those days.

By travelling to farms throughout the county, Henry, utilising the experience and knowledge gained from his early years with the Hull Cart and Wagon Company, also recognised the problems that transporting goods

Fig 143
Advertisement from Peacock & Binnington, c1910
(Note the horse-gear in the foreground).

to and from farms to railway stations created, especially for the Fenland farmers of South Lincolnshire, which were often situated some distance from the nearest road or railway. Hence Peacock and Binnington became agents for the sale of many miles of narrow gauge railway lines to these farmers. It was claimed in a newspaper report on Spalding's light agricultural railways that, with this system installed on a farm, one horse could move up to seventy tons of produce per mile per day, a figure vastly above what could be achieved by normal methods.

Not just content to concentrate on implements and machinery, however, Peacock and Binnington were also involved in the installation of telephones into remote farm houses at a time of limited communication facilities. They also supplied farmers with acetylene lighting sets, and it is interesting to see an early twentieth century photograph of Peacock and Binnington's stand at the county show carrying an advertisement for The Leading Light Syndicate Acetylene Gas Engineers, Hull. This would seem to confirm that Peacock and Binnington acted as agents for this local company.

Fig 142
An important early selling line for Peacock & Binnington.

Fig 144
Display of machinery sold by Peacock & Binnington lined up in Bridge Street, Brigg (date unknown).

Peacock and Binnington were now beginning to get wide recognition. The November 1911 edition of the *Implement and Machinery Review* stated that: 'while as an ardent advocate of the use of all kinds of labour-saving machinery on the farm, estate and dairy, Mr. Peacock has worked hard and travelled much, till now his firm owns one of the largest retail implement businesses in the East Riding of Yorkshire and the Eastern Counties.' Mr Henry Peacock himself observed that: 'All I can say is that when we started in the implement business thirteen years ago, we had no connection and now we have. We started unknown.'[8]

World War I and Changes at the Firm

However, dark events were gathering at pace in Europe and soon the world was to be plunged into the Great War. Away from the front lines British agriculture was also going through changes. Before the war Britain was importing much cheap food, which adversely affected the cost of home grown produce, much to the financial detriment of British farmers and those companies which supplied them with machinery. The war meant that the importation of food into the country decreased substantially as vessels were in greater need for moving troops and armaments. Farmers were finally beginning to get what they believed was a proper price for their goods and, to the benefit of agricultural implement suppliers like Peacock and Binnington, the increased farm productivity meant a greater demand for farm implements and spares. Unfortunately, this new found optimism in the industry would not last long and after the war cheap imports began again and the depression in British farming continued into the 1930s. Prices for home-produced food fell sharply as did farm workers' wages. It was a struggle for survival, but Peacock and Binnington again managed to continue trading during this bleak period and emerged from it on solid foundations.

It was during this period that, in 1921, John Binnington, co-founder of the firm, made the decision to retire. By that time, however, he had left most of the operating of the business to Henry Peacock. Little is actually known about John Binnington or his contribution to the business other than that he provided the main financial backing to the company. It is believed that he never once visited the Brigg depot. His decision to retire also meant the effective end of the Hull connection, because Henry Peacock decided to run the business solely from Brigg. This may have been of some relief to Henry as the only method of transport between the two depots was the time-consuming ferry and 'he later calculated that in total he had spent nearly a year of his life on the Humber Ferry.'[9] It was by strange coincidence that the Hull depot was sold to a man also named John Binnington, no relation to the other one. However, such was the recognised strength of the company name that it was never changed, and Binnington still remains in the title some eighty years after his retirement.

When in 1937 Peacock and Binnington became a 'limited' company, Henry Peacock was in charge as managing director, with his fellow directors being second son, Dennis, wife Mary and company salesman, Ben Benstead. Henry's eldest son and future company president, Henry Allen, was installed as the company secretary. However, Henry Earl Cartwright Peacock died in 1938 before the limited company's first annual general meeting. An indication of the impact that Henry had had on Lincolnshire agriculture was shown by the great many tributes that poured in from all those that knew him. The respect from his peers could be summed up by a Mr J L Goldsmith from Ipswich who wrote, 'The implement trade generally will be the poorer for the loss of one who always gave of his best to keep the trade on a high plane through his personality and wonderful service.'

The job of managing director of Peacock and Binnington passed to his son H Allen Peacock, who, educated at Brigg Grammar School and Lincoln Technical College, had served engineering apprenticeships at both Rustons of Lincoln and Ransomes, Sims and Jefferies of Ipswich. When he first joined Peacock and Binnington in January 1922, only a few months after John Binnington had retired, it was very much the workshop side that he was attracted to and he mainly left the selling to his father. Peacock and Binnington were at this time instrumental in the selling of binders, ploughs and steam ploughing sets. Yet Allen still continued with his father's principle of service before sales. The philosophy was that, by providing the farmer with excellent back up on machinery already sold, the customer would more than likely return to Peacock and Binnington when the machine needed replacing. Allen Peacock remained the president of the company until his death in 1985, aged eighty-six.

World War II and the Post-War Years

Unfortunately, shortly after Allen Peacock took over as managing director, Britain again found itself at war. Like every other aspect of normal life, the agricultural industry faced its own problems when the supply of machinery and implements was taken away from the

Fig 145
George Bradshaw of Peacock & Binnington offering tuition to a female Land Army tractor driver, 1940s.

Fig 146
Display stand at the Brigg Industries fair in Brigg Corn Exchange, 1950.

free market and placed in the hands of the War Agricultural Committees. In effect these committees dictated how agricultural supplies were to be allocated and the resulting bureaucracy meant delays and restrictions in the selling of machinery by dealers such as Peacock and Binnington. Following the agricultural depression during the 1920s and 30s, this was just another burden imposed upon an already struggling industry. British factories were now mainly given over to war supplies and much agricultural machinery was imported, often through the war-time 'Lease-Lend Scheme', mainly from the United States and Canada. Machines such as the Massey-Harris 21 combine and No.15 trailed combine arrived in great numbers, both during and for some time after the war. Packed in kit form into space-saving wooden crates, consignments of these machines would arrive on railway wagons at Brigg station and be unloaded by crane. Peacock and Binnington engineers would assemble them on land next to the station before driving them through town to the Old Foundry site.

With so many men involved in the war effort, labour was in short supply and Peacock and Binnington relied heavily on the skills of long-term employees such as Ben Benstead and George Bradshaw for the sales and service side of the business. Allen Peacock managed to look after the Brigg office, whilst combining his time with the Royal Observer Corps at nearby Scawby Moor. Time was also spent training the Women's Land Army in the use of farm machinery that had traditionally been the work of the men *(Fig 145)*. The shortage in man-power during the war years was somewhat alleviated by the use of prisoners of war based at nearby camps, including Pingley Camp, which was situated close to the eastern outskirts of Brigg.

After the war the agricultural economy gradually improved and there was a rapid rise in large-scale farm machinery. Peacock and Binnington's business began to grow and expand. In 1958 a combine repair shop was built at the rear of the Old Foundry premises on land that was previously used as a rifle range.

Fig 147
Michael Peacock and Henry Allen Peacock.
(grandson and son of the founder).

1968 saw the company purchase the Old Foundry site that had always been leased from the Earl of Yarborough and this allowed old, mainly wooden, buildings to be demolished and replaced by new modern premises. A new office block, designed by local architect John Thompson, was erected in 1975, with Allen Peacock given the honour of laying the commemorative stone. In 1963 George Bradshaw was rewarded for his considerable efforts by becoming a director. It was in that same year that current owner of the company, Michael Allen Peacock, Henry Earl Cartwright Peacock's grandson, joined the board *(Fig 147)*. Michael, born in 1933, was educated at Brigg Grammar School before spending eighteen months familiarising himself with agricultural engineering, as did his father, with Ransomes, Sims and Jefferies. He later gained an engineering degree at Emmauel College, Cambridge.

Peacock and Binnington had already extended their business by opening a depot at Laceby in 1929 and in 1952 they took over a blacksmith's business belonging to J Fletcher at Market Rasen. Since 1959 they have been associated with Massey-Ferguson under a dealer arrangement that meant that Peacock and Binnington would not actively promote any other make of tractor. However, through tough negotiation the firm maintained the right to sell other manufacturers' machines if the customer insisted. Michael Peacock recalled that 'at that time we tended to supply whatever the customer wanted and my father took a stand of not being dictated to.' However, other machines gradually diminished from the company's showrooms as Massey-Ferguson became the major tractor franchise, which it still continues to be today.

In 1970 Peacock and Binnington continued their expansion with the taking over of the 'Greenacres' Massey-Ferguson dealership premises in Louth. In 1980 a new building was erected on the Louth site which later became a showroom and workshop that enabled the branch to expand into the car business. As dealership territories continued to change, the Laceby depot was closed in 1989, followed by Market Rasen several years later. However, new depots were opened at Corringham in 1985 and Hatfield, South Yorkshire, in 1993 as new trading areas were recognised to the west of Brigg.

Currently plans are emerging for the two smaller depots at Hatfield and Crockey Hill, York, to be amalgamated to form a new depot at Selby, thus widening Peacock and Binnington's influence in the Yorkshire area. Any business depends on the strength of the product that is being sold and over the past few years Peacock and Binnington has built up a strong collection of franchises, including Massey-Ferguson, Challenger and Fendt tractors, and implements made by manufacturers such as Kuhn, Welger and Simba, together with the range of JCB agricultural machinery.

Fig 148
Massey-Ferguson balers at Brigg station awaiting delivery to Peacock & Binnington, 1968.

Fig 149
Fleet of Massey-Harris combine harvesters ready for harvest, c1965.

Peacock and Binnington remains an active business in the Lincolnshire farming industry after more than 110 years. Its company chairman, Michael Peacock, grandson of the founder, has recently held the honorary position of President of the Lincolnshire Agricultural Society, a mark of the esteem in which he and his company are held within the county.

The company, carrying an estimated 30,000 part numbers in stock and covering dozens of franchises, still proudly maintains its philosophy: 'service first, then sales'.

Notes

[1] Cook, *Hull Alphabetical Directory*, 1897
[2] *Kelly's Directory of Lincolnshire*, 1913
[3] *Kelly's Directory of Lincolnshire*, 1900
[4] *Lindsey and Lincolnshire Times*, 19 September 1896
[5] *Farming Echo*, August 1994
[6] *Dewsbury News and Chronicle*, 23 June 1900
[7] Jackson's *Brigg Annual*, 1907
[8] *The Implement & Machinery Review*, 1 November 1911
[9] *Farming Echo*, August 1994

Fig 150
Peacock & Binnington's stand at the 1994 Lincolnshire Show, still selling Massey-Ferguson products.

CHAPTER 12

JOHN H RUNDLE OF NEW BOLINGBROKE

IRONFOUNDER AND FAIRGROUND RIDE MAKER

Alan Rundle

Haulage and Steam Threshing

John Harness Rundle started a haulage business in 1913 at the age of seventeen in New Bolingbroke, ten miles north of Boston. Although he was the son of the village vicar, he had always been interested in mechanical things, having had a lathe from an early age. His engineering interest probably went back to his great-grandfather, who was a partner in Gill & Rundle's Foundry at Tavistock in Devon. Here they made beam engines for the local mining industry, but this company closed with the demise of the copper mines in the mid-nineteenth century.

In October 1913 John Rundle with his father, the Reverend Harness Rundle, travelled to Lincoln to visit William Foster & Co Limited. There they ordered a new 5-ton Wellington steam tractor complete with trailer *(Fig 152)*. On the 14 January 1914 John returned to Lincoln to collect his newly built engine and was surprised to find that Fosters had named the engine 'Venture' in recognition of the seventeen-year-old's business investment.

Initially he operated from the Vicarage in New Bolingbroke, but later, in 1915, he married Bertha Twelvetrees from The Shop in New Bolingbroke Crescent and moved to Mill House and yard at the north end of the village *(Fig 151)*. This is where the business was built up and remains to this day. The only buildings on the site at that time were the house, a disused Methodist chapel, a bakehouse and the windmill tower. The steam tractor was kept busy because it was capable of carrying a greater capacity than the horse-drawn carts of that period. One incident at Holmes of Wragby (timber merchants) illustrates the sort of work they were asked to do. A customer wanted to put some more timber on Rundle's over-laden trailer but when it was taken to the weighbridge the five-ton trailer was already carrying over fifteen tons!

Half-way through the First World War the engine was commandeered for military use (mainly haulage work, but also steam sawing) and John was forced to join up to stay with the engine. After the war he reverted to his

Fig 151
The Mill Yard, New Bolingbroke, as purchased by John H Rundle in 1915.

Fig 152
Rundle's original Foster steam tractor, purchased in 1914.

haulage work, much of which involved leading road stone from the railway and piling it in heaps on the roadside for road repairs. The council contract paid two shillings and sixpence a ton and involved digging the stone out of deep railway trucks by hand.

By the early 1920s the days of the steam tractor were numbered, as the petrol lorries used in the war were being sold off and were beginning to displace steam power. In 1922 he went back to Fosters at Lincoln and purchased parts to convert his engine to a showman's

Fig 153
Threshing at New Bolingbroke with Robey traction engine, 1929.
John Rundle senior is standing by the wheel; his son John junior is on the footplate.

Fig 154
The sale of threshing sets at New Bolingbroke, 1947.

type. From 1923 John travelled the fairs generating power for the rides and hauling them from site to site. But this was not the life his wife wanted because by now they had a young son (born in 1921), also called John, and spending the summers living in a caravan did not appeal. So after four or five years he returned to general haulage and bought a Burrell steam wagon, followed by a traction engine and a threshing machine. The Burrell wagon was not a success and, despite the manufacturer coming out to look at it, it never ran well. It was only in later years when scrapped that the cylinder was found to be a faulty casting. In any case, by the early 1930s, steam wagons were taxed off the road. This left them with the threshing business and soon more engines and threshers were purchased. The traction engines were bought second-hand, but the threshing machines were new from Fosters.

In this period George Gosling, better known as 'Gander', was asked to repair a drum axle; he was taken on full-time and worked for Rundles for the following sixty years as an engine driver, carpenter, pattern maker and finally looking after the engines at steam rallies. The third item in a threshing set is the straw elevator, and those available at the time were not ideal, so in 1935 John designed and built a new elevator. In the following years he built more elevators for his own use, making seven in total by 1946.

John Harness Rundle junior joined the firm on leaving school in 1936. He was always more interested in the engineering side of the business rather than threshing. By now, besides servicing and repairing their own equipment, Rundles were repairing those of other owners and contractors.

After the Second World War the arrival of diesel tractors and combine harvesters marked the beginning of the end for steam threshing. So in 1947 the six sets of threshing tackle, all complete with a Rundle elevator, were auctioned off *(Fig 154)*. Some of the traction engines were bought back in order to save them from the scrapman's torch, but this was the end of the days for Rundles as threshing contractors. With hindsight the money from the auction ought to have been invested in a better workshop and new machine tools; this was a missed opportunity.

John junior's ambition was to start a foundry to enable them to produce their own iron castings. When Stephenson's foundry closed in Tower Road, Boston, the manager, Len McDermot was approached with the view of establishing a foundry at New Bolingbroke. Around 1950 a small cupola furnace was purchased and castings were produced for the first time at the side of the mill tower. Unfortunately, before the foundry really got off the ground, McDermot died suddenly, so although the foundry was partly up and running, they were only making parts for their own use.

Up to the sale in 1947 all the elevators built had been for their own use, but afterwards they began to sell them to farmers and contactors. In 1950 the Lincolnshire Show was held at Spilsby and the Rundle Elevator was exhibited for the first time, resulting in four elevators being sold. Rundles went on to exhibit at the Lincolnshire Show for the next forty-five years. The elevator was a popular machine; it had a thirty-foot long trough, each side being made from a single plank of timber, and it was light and manoeuvrable in the stack yard *(Fig 156)*. As with threshing by steam, threshing by tractor was also being taken over by the combine harvester and with it went the sales of straw elevators. The last new elevator was built in 1960, the most successful year being 1955 when seven were built.

The Globe Foundry

In 1952 the Globe Inn in New Bolingbroke closed and was then sold by auction. The inn, located three hundred yards to the south of the Mill Yard on the east side of the main road, was bought by the Rundles with a view to converting the old brewery buildings into a foundry. John junior was by now married with a family of his own and the front of the Globe Inn was converted into his family house and the back into another house for the foundry manager *(Fig 155)*. The name Globe Foundry was adopted and many of the castings had the name 'Globe' on them as a trademark.

Most of the cast-iron scrap was bought locally and usually had to be collected. Along with the cast-iron, there was often as much, if not more, steel scrap. This was before the days of skip-lorries, and the steel had to be disposed of in a railway truck at New Bolingbroke

Fig 155
The Globe Foundry, New Bolingbroke, 1954.

railway station, from where it was transported directly to Sheffield. Likewise, the coke for the furnace was delivered to the station and it was then a day's job for the foundry men to throw the large lumps of coke into a lorry before it could be moved to the foundry, where they would have to shovel it out again. The castings were also dispatched from the station up to the closure of the railway line in 1970, by which time British Road Service was operating an alternative parcel service.

In 1955 the firm of James Coultas & Co from Grantham closed down and the patterns and spares for their corn drills and rising hopper distributors were bought. A few of the distributors were built at Rundles, but most of the sales were in spares, the last of these being sold around 1970. Hornsby of Grantham had also been manufacturers of seed drills and by the mid-1950s

Fig 156
The Rundle straw elevator made in the 1950s.

*Fig 157
Moulding in the foundry
(Jim Powell and Colin Clayton),
1956.*

they realised that these did not fit into their current product range. The patterns and spares were bought by Rundles and, as with James Coultas, new parts were supplied for Hornsby drills across the country for many years. Further afield, Richmond & Chandler of Manchester closed down in the late 1950s. Another firm of international reputation, they had produced bakery equipment, chaff cutters and friction hoists, and once again the patterns and the spares were purchased. A number of the friction hoists were built, the first in 1959, and this was probably one of the earliest uses of electric motors on new machines at Rundles.

In the late 1950s corn began to be stored in bulk rather than in the traditional hessian sacks and, in order to keep it dry and prevent it from over heating, large fans were used. John Rundle junior saw a future demand for fans in the corn drying process and Rundles was one of the first firms in this market. Most of the fans were second-hand ventilating units bought in the Midlands and then refurbished before installing on the farms. Fifty years on, corn drying fans are still an important part of the business.

At the 1958 Lincolnshire Show William Foster & Company Limited of Lincoln exhibited their products on John H Rundle's trade stand and in 1961 they approached John Rundle with a view to selling the agricultural side of their business. Since the sale of the threshing business in 1947, a large part of the engineering work had been the servicing and repair of 'threshers'. Rundles was the obvious customer for the agricultural side and Fosters were later to change their name to Gwynnes, concentrating on the manufacturing of pumps.

For the sum of £1,250 Rundles received three incomplete threshers, many tons of spares, besides the hundreds of patterns for chaff cutters and threshers. The firm's Foden lorry was kept busy for weeks running the 30 miles to Lincoln and back. This lorry, with its Garner 4LK engine, could manage 32 mph on the flat. On one trip carrying a threshing machine and all the wood that could be threaded underneath, Harold Squires rode all the way up Canwick Hill on the edge of Lincoln with his door open ready to bale out. Maurice Roberts, the driver, said Harold only closed it on struggling to the top. In September of the same year, 1961, one of the part-built threshing machines was completed and sold to A P Clarke of Little Carlton near Louth. This machine, number 9089, was the last new thresher of any make sold in the country *(Fig 158)*.

In order to accommodate the Foster drawings and records at New Bolingbroke a new office was built on the top of the flat roof of the engine house.
Mr Stamper from Lincoln was employed to look after the enquiries and sales. He travelled to work daily by train to New Bolingbroke station and strolled through the village to Rundles wearing his trilby hat and carrying his briefcase and umbrella. This was quite a sight for a small rural village in 1961!

*Fig 158
The last Foster threshing machine to leave the works, 1961.*

Fig 159
Pig troughs made by Rundles on display at the Lincolnshire Show.

In 1962 the first full year of trading in Foster spares showed a turnover of £3,423.1s.1d. This was boosted by thirty sets of safety guards to cover the belts to comply with the new regulations. Two years later spares sales had dropped to £1,006.9s.8d, reflecting the decline of threshing, but parts continued to be supplied up to 1980.

Another acquisition in the 1960s was the 'Goodwill of the Maynard Chaff Cutter and the Master Mould and Patterns, together with some spare parts as available for the sum of Fifty-five pounds.' Robert Maynard had recently died and his stock and equipment was sold by auction at the Whittlesford Works, near Cambridge, on 16 March 1961.

With every additional line of spares came new customers and by 1962 most of the counties in England, Scotland and Wales were being supplied. This helped Rundles expand beyond the county to a nationwide customer base. Since the foundry had started there had been two changes of manager, Joe Powell taking over from Arthur Tinkler in 1958 and then Ray Picker taking over in 1963. Ray continued as foundry manager until the closure in 2003 and, apart from doing two years National Service, had been employed by the company for over fifty years.

The Limited Company Established

When Rundles became a private limited company (John H Rundle Limited) in 1965, the two main products of the foundry were pig troughs *(Fig 159)* and Cambridge roll rings. Pig troughs and cattle pans were produced in larger quantities until 1964, when for the first time more roll rings were made. At this time the total tonnage of all castings for a year was 115 tons. The Cambridge roller, a simple agricultural implement for crushing clods, had been popular for a hundred years, being invented by William Cambridge of Market Lavinton, Wiltshire. It consisted of cast-iron spoked rings with a curved v-section on the rim, usually three inches (7.5cm) wide and about twenty-four inches (30cm) diameter. These rings were slid onto an axle to make a wide roller. The early rollers were six feet (1.8m) wide or less, but as with all farm implements they became wider and wider. Eventually a rolling width of forty feet (12m) was achieved by using five separate rolls in a gang roll.

Up to 1969 one hundred roll rings were considered a good order until R Hunt & Co Ltd of Earls Colne placed a single order for five thousand rings. In the same year orders were placed as follows: Edlingtons of Gainsborough - sixteen hundred rings; E J Tong of Spilsby - five hundred machine castings; Robey of Lincoln - twelve hundred castings; Yannedis & Co Ltd of London - six hundred and fifty sash weights; and W Sugden of Barking, London - eight hundred electric motor slide rails.

By now the workforce at the foundry had increased to ten and at its busiest in the 1980s it reached fourteen men. In 1973 the first kerb offlets (drainage gulleys placed between kerbstones on the roadside) were made for Lincolnshire County Council. Over one thousand a year were produced for the next thirty years, and they can be seen on the roadside all over Lincolnshire and in parts of Norfolk and Nottinghamshire too *(Fig 160)*.

Fig 160
Kerb offlet in Dunholme, one of thousands cast by Rundles.

Around 1980 another old farm implement had a revival, this being the furrow press. Comprising two rows of heavy, sharp-pointed rings, it was usually trailed behind a plough to firm the soil *(Fig 163)*. There were several patterns of furrow press wheels, some weighing up to 150 kg each. With roll rings and furrow press wheels, the foundry was casting every day, and in 1987 over fifteen thousand castings were made, weighing a total of 251 tonnes.

The old foundry became too small to cope and in 1988 work started on a new building, eighty feet long by forty feet wide (24 m x 12 m) on the south side of the old one. In May 1989 production moved to the new foundry and in the first year it turned out 237 tons of castings. However, in 1991 one of the best customers, Cousins of Emneth, which had been buying 4,000 rings annually, started importing Indian castings at a much lower price. That year the total tonnage cast dropped to 141 tonnes.

Castings from both India and China were taking all the large volume markets, leaving the Globe Foundry with small quantities and special one-offs. By the year 2000 the number of castings had dropped to 4,037, with a total weight of seventy-nine tonnes. The workforce had gradually reduced to three full time men and one part-time, so that when casting in the final year, two men were sent from the engineering yard each afternoon to help. When Ray Picker, the manager, reached retirement age in 2003, the difficult decision was made to close the foundry after fifty years of trading. Ray continued to work part time, the small brass foundry being retained for aluminium and brass castings. In 2004 the main foundry and pattern stores were let to PGM, a fibreglass moulding company.

When Rundles bought the drill spares from James Coultas and Hornsby they also required sheet metal shutes and slides, which were made by C A May the tinsmith at Boston. When Charlie May died in the mid-1960s, Rundles employed a tinsmith and carried on the business. The tinsmith was originally situated at Bargate Bridge, but when the bridge was widened they moved to 27 Pipe Office Lane off West Street. In about 1972 this site was made into car parking and C A May moved back to a site behind Holland House near Bargate Bridge. In 1974 this latest site was to be demolished to make room for John Adams Way relief road and this time the business closed and all future sheet metal work was carried out at New Bolingbroke.

Fig 161
Pouring the moulds in the foundry.

Fig 162
The foundry's coke-fired cupola furnace.

In the 1960s the pea vining industry provided valuable work for Rundles in the manufacturing of pea tanks and conveyors as well as in carrying out essential repairs. This all changed when mobile pea viners took

Fig 163
Hydraulic folding furrow press, designed and made to order by Rundles.

over from the static machines and the peas went straight from the field to the factory, ending the need for pea tanks and conveyors on the farms. Rundles also carried out repairs on the early, unreliable drainage machines for the drainage contractor Fred Coupland of Carrington. When John Mastenbroek from Holland first left Coupland to set up on his own, his men used the workshops at New Bolingbroke and later built their own premises at Boston.

At the mill yard the main workshop was increased in stages until it surrounded the mill tower and reached the full length of the northern boundary. A sectional building, measuring forty feet square, was bought from East Kirkby airfield and was erected in the middle of the yard and used as a showroom for selling electric motors and machine tools. Since 1943 The Maltings at New Bolingbroke had been rented for storage, but in 1966 the owners decided they needed it for themselves. Measuring over one hundred feet (30 m) long by twenty-five feet (7.5 m) span, it had stored a lot of equipment and to accommodate this, a Second World War Blister hangar was erected to the south of the mill property. In 1972 another building was erected to the back of the hangar and half of this shed was to be a museum. The Rundle family had always valued their heritage and in the museum building they housed three steam engines, the Foster tractor bought new in 1914, the unique Brown & May showman's engine bought in 1936 and a Savage centre engine owned since 1950.

From the 1950s John Rundle junior had been selling used electric motors and, as farms became more mechanised, this market grew rapidly with many motors being fitted to corn augers and drying fans. In 1968 Jack Rundle secured the sales of MEZ electric motors, made in Czechoslovakia, by becoming their official stockist. By 1980 Alan Rundle was running the electric motor sales, often with as many as four thousand new and second-hand motors in stock.

John H Rundle the founder of the company died in 1974, having worked well into his seventies. He was survived by John H Rundle junior and his four children, Jack, Ken, Sheila and Alan. Jack, the eldest, joined the company in 1966 and for many years ran the machine tool sales side of the business, but left in 2003 to concentrate on his antique business, 'Junction'.

Ken was the last to join the family firm; having first studied for his Degree in Engineering at Loughborough, he then went to work for Clayton Dewandre at Lincoln developing anti-skid brakes. After the death of his father in 1990, Ken became the Managing Director and now looks after the fairground and manufacturing side. Sheila joined the firm in 1970 to work in the offices and is also the company secretary. The youngest brother, Alan, started work in 1972 and took over the electrical side of the business; he also maintains and exhibits the family's steam engines.

During the 1970s and 80s the engineering works produced many complete sets of Cambridge rollers. At the busiest, a roll would be completed every day and was usually delivered with Rundles' own lorry. A number of rolls were exported, going as far afield as Egypt, Canada and the Falklands. Special rolls were also built to the farmers' own specifications and many of the later machines were hydraulic folding ones.

Fairground Rides

In 1968 Butlins Holiday Camps approached Rundles with a view to rebuilding a fairground ride from their Ayr camp in Scotland. This was a set of galloping horses that had to be taken off the old wooden centre truck and mounted on a permanent base. In the following years Butlins sent six more sets of gallopers from their other camps around the country for major rebuilds. Several other rides have been repaired for Butlins since then, including Waltzers, Noah's Arks, Derby Racers, Ferris Wheels and Miniature Railways. Two new monorail trains have been made for the Skegness Camp and in recent years two road trains have been built for that site.

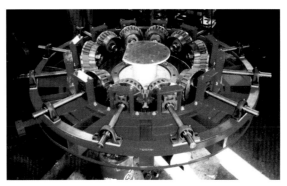

Fig 164
Twelve-section galloper centre for fairground ride.

The galloper was developed in the late nineteenth century, the last British-built machine being made in 1924. Some of these rides are now well over one hundred years old and are becoming difficult to repair satisfactorily. One such ride, a Walker-built machine from Butlins at Pwllheli, had to have a complete new centre made *(Fig 164)*.

In 1991 work started on building a completely new set of gallopers; this was to be a fourteen-section, four-abreast machine of fifty-feet diameter, equal in

Fig 165
Four-abreast gallopers fairground ride built for Loudon Castle, Scotland.

Fig 166
Leisure road train constructed by Rundles and powered by Massey-Ferguson turbo-diesel engine.

size to the largest machine built by Savages of Kings Lynn. Work on this new set only took place when the workshop was quiet and it was 1995 before it was completed. It was then leased to Loudoun Castle near Kilmarnock in Scotland *(Fig 165)*. The Victorian ride on which it was based had needed daily maintenance, but the new machine used self-lubricating plastic bearings and other modern materials, making it virtually maintenance free. Instead of a wood top frame, steel was used and all the horses and rounding boards were moulded in fibre glass; the platform and steps were fabricated from aluminium with wood tops. Like the manufacturing one hundred years earlier, everything was made on site. This included over two hundred castings from the foundry, all the machining and fabrication work, the only exception being the electric motor, gearbox and electronic controller that were bought in.

To date there are seventeen sets of gallopers with Rundle centres, seven of these being complete new machines, the others being rebuilds. The last old ride to have a complete new centre was for Blackpool Pleasure Beach, and Rundles are currently building a new ten-section three-abreast galloper for a customer on the south coast.

Besides the gallopers, other fairground rides have been repaired and refurbished, including modern thrill rides, such as Top Spins, Terminators and Super Bobs. Wicksteed Park, near Kettering, has bought a set of Gallopers, a Seal Ride and a Veteran Car Ride from the company, as well as having the Water Chute rebuilt and a new drive system fitted to the Twin Pirate Ships.

A branch was opened in Scunthorpe in 1992 called Rundle Electric Motors in a bid to expand the business in this industrial town. However, this was closed in 1994 when the company bought a controlling share in Lincolnshire Rewinds at Lincoln. Lincolnshire Rewinds had been started in 1968 by Brian Skinner and Dave Rowley, two ex-employees of Penney & Porter Ltd. This was the only electric motor rewind company in Lincoln, and in 1992 employed six people. Brian the senior partner retired, leaving Dave running the business in partnership with John H Rundle Ltd. Lincolnshire Rewinds specialise in motor rewinds and pump repairs and are the official stockists of Brook Crompton products.

Today, international sales play an increasing part in the company's business, aided by a comprehensive website.

CHAPTER 13

TUXFORD & SONS OF BOSTON
STEAM ENGINE BUILDERS

Neil Wright

Tuxford and Sons has been credited with the invention of the steam threshing sets that were in use on British farms for over a hundred years until combine harvesters took over *(Fig 170)*. This engineering firm lasted less than fifty years and probably never had more than 200 employees but it sent steam engines worldwide. Some are still in museums in Paris, Sweden, Edinburgh and Lincoln, and in private ownership in Australia and England.[1]

William Wedd Tuxford, Founder of the Firm

The firm began in the 1837-40 period, when William Wedd Tuxford (1782-1871) was joined by his four sons. Tuxford had started in business as a miller and baker over 20 years earlier and at that time had no idea of starting an engineering works.[2] By 1822, if not earlier, he had an eight-sail windmill in Skirbeck, next to Mount Bridge, just outside the boundary of Boston borough. His baker shop was in Boston Market Place at what is now No.21, and he lived there for much of his life *(Fig 167)*. He came from a large family who were active in several aspects of Boston life in the nineteenth century. When William's father, Weston Tuxford, died on 25 July 1837, it was said that he had eight children, 53 grandchildren, 49 great-grandchildren and one great-great-grandchild.[3] Those eight children included William's brother, James Edward Tuxford (1774-1855), described as a 'pawnbroker' but really something of a small scale banker, prepared to invest in different projects at different times. During the next half century members of the Tuxford family were to be one source of funds for William's engineering business.

It is curious that William Wedd Tuxford's windmill was built outside Boston, since at the time there were still many green field sites within the borough. The reason he chose a site outside town might have been political, as William was a radical with revolutionary and democratic principles. In the early nineteenth century Boston Corporation was controlled by a powerful Tory clique that was to be totally overthrown after the passing of the Great Reform Bill and the Municipal Corporations Act, and Tuxford was one of the radical councillors elected in December 1835.

In the early 1800s two Boston bankers, Henry Clarke and Henry Gee, bought the six-acre Church Pasture in Skirbeck, between the Maud Foster Drain on the west, the Haven Bank on the south, the grounds of Skirbeck rectory on the east and Fishtoft Road on the north. They divided the large field into smaller plots and

Fig 167
21 Market Place, Boston.
William Wedd Tuxford lived above his baker's shop here for much of the 19th century (photograph 2005).

Fig 168
Extract from painting dated 1822 showing Tuxford's original works from the south; Boston Haven and Skirbeck Sluice in the foreground.

started selling them. By December 1822 Tuxford had purchased a site nearest Mount Bridge, containing about three-quarters of an acre *(Fig 169)*. Three other smaller sites were sold east of Tuxfords and fronting Fishtoft Road, but then the Reverend William Roy, the rich new Rector of Skirbeck, bought the rest of the Pasture and prevented any further development encroaching on his grounds.[4]

To finance his purchase of land at Mount Bridge, Tuxford borrowed £2,000 from his brother Peter Tuxford, secured by a mortgage dated 31 December 1822, in which the property was described as 'All that Wind Corn Mill, Machine House, Dovecote, Stable, Carthouse, Cowhouse and two pieces or parcels of land containing by admeasurement 3,534 square yards'.[5] A painting dated August 1822 shows the eight-sail windmill and other buildings clustered around it *(Fig 168)*. In 1856 it was said that the works were thirty years old, but that refers to the decade when Tuxford invented a reeing machine; his business did not develop into the Boston and Skirbeck Ironworks until the late 1830s.

From Miller to Machine Maker

During the wet summer of 1824 William Wedd Tuxford, the miller and baker, went to great trouble to separate the sprouted wheat by hand, and hence his flour made

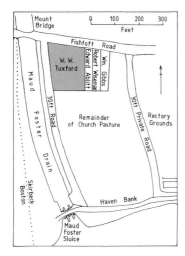

Fig 169
Location of Tuxford's original works on the Church Close.

10d per stone more than any other on sale at Boston market. He then considered how the same process could by done by machine and after much hard work invented a 'double motion reeing sieve' and patented his design (Patent No.5954) in 1830. Tuxford asked a local craftsman to make the castings for his machine, but this work was refused, so Tuxford made them himself. The demand for these successful reeing machines was so great that their production became a separate activity in the workshop beside his mill and was the genesis, ten years or so later, of the Boston and Skirbeck Ironworks. As late as 1870 a picture of the reeing machine, well worn by time, still held pride of place in the Skirbeck works' office. With a variation in the gauge of the wire mesh it had also been used for riddling grass seeds, linseed and coffee beans, and sent to Egypt for lentils.[6]

In the 1820s and 1830s Tuxford was listed in trade directories as a miller, baker and flour dealer, and in 1835 he also described himself as 'patentee of reeing machines'. The directories had categories for 'iron and brass founders' and 'engineers' but Tuxford did not include his business in these until 1840. The 1822 mortgage referred to one building on the site as 'Machine House', so that was perhaps the workshop where he developed and then manufactured his machines. In 1870 it was said that Thomas Sampson and Ellis Maddison had both worked for Tuxford for forty years, so they must have worked in the reeing machine workshop for the first few years. When the works was extended in 1859, the plan showed the buildings on the original site grouped around a central yard with the windmill (W) in the south-east corner *(Fig 171)*.

It was in 1837-40 that the small-scale machine shop next to the windmill grew into the Boston and Skirbeck Ironworks. The earliest newspaper mention of Tuxford is on 18 August 1837, when the *Stamford Mercury* reported that in Spalding the 'new Victoria iron suspension bridge (cast by Tuxford and Son of Boston) is in rapid progress, and, judging from present appearances, will have a beautiful effect and be highly ornamental to this part of the town'[7], so they were making something much larger than a reeing machine. It was also in 1837 that William Wedd Tuxford borrowed £2,000 from the Stamford, Spalding and Boston Banking Company in addition to his 1822 mortgage, which supports the other evidence of an expanding business. By 1838 two of his sons had joined the firm, Weston (1814-85) and Wedd

Fig 170
Tuxford steam threshing set of 1871 in use on the Thorold estate at Syston, near Grantham, up to 1914.

(1806-94), followed by Joseph Shepherd (1808-78) and William (1820-post 1882) by 1842. From this time until its end the firm was known as Tuxford and Sons.[8] At their peak they were a partnership of a father and his four sons. It would be fascinating to know how the dynamics of that worked, but alas we have no company records or personal diaries to help us.

All four sons joined the firm at about the same time and it seems that Weston may have been the most innovative and important in the development of the firm; for its last five years he ran it as the sole partner. Weston's obituary in 1885 says that his father built the windmill and patented the reeing machine, but then Weston, 'who had been educated as an engineer in the neighbourhood of Grantham, came and took charge of this branch of the business'. This may mean that Weston was apprenticed in Richard Hornsby's Spittlegate Ironworks on the edge of Grantham before joining his father in Skirbeck to transform the machine-making side into an engineering works. Weston's obituary says he was 'an engineer of great ability, and a large proportion of the improved agricultural machinery of the last half century has been produced from his inventions.'[9] It particularly credits Weston with the development of the threshing machine, and other sources link Weston's name to patents in 1850 and 1854 for improvements to portable steam-engines, threshing machinery and clod-crushers.[10] In 1874 it was said of Joseph Shepherd Tuxford that he was 'well versed in agricultural mechanics' and understood principles as well as mechanical details. When Joseph died in 1878 he was referred to as the 'senior partner', but that may simply mean that he was older than his brother Weston.[11]

Steam Engines and Threshing Machines

The firm expanded into boiler-making and the *Stamford Mercury* of 12 June 1840 reported the explosion of the Tuxford-made boiler of the steam packet boat 'XL' that regularly traded on the river Witham between Boston and Lincoln.[12] About this time the firm started exhibiting at the annual shows of the Royal Agricultural Society of England and its successes there led to orders from customers in colonies of the British Empire and other foreign countries, as well as from Britain. Tuxford & Sons was rapidly taking its place as one of the firms which were making Britain the workshop of the world.

In the late 1830s several rural foundries were exploring the application of steam power to agriculture. In 1839 William Howden of Boston produced the first steam engine to be made in Lincolnshire and went on to make the first portable threshing machine in 1841. In 1839 it was suggested to Tuxfords that they design and make a portable steam-powered threshing machine. A design and working models were produced, but they were relegated to a shelf in the works' office until 1842 when a full-size prototype was completed *(Fig 172)*.[13] Did this three-year delay reflect that W W Tuxford was more cautious about new ideas than his sons, even though they were now his partners? The Tuxford design went a stage further than Howden and combined the steam engine and threshing machine on one frame or chassis, the whole being carried on four road-wheels and with shafts for horse haulage. It was not a complete success and in the autumn Tuxfords separated the two components and produced the first combination of a portable steam threshing set. This included, in their simplest form, the elements of the threshing sets which were in use on British farms until the 1950s, when combine harvesters took over *(Fig 170)*. Tuxfords' steam threshing machines got good coverage in the *Stamford Mercury* and the firm sold several to local farmers during 1843.[14]

Then a disaster struck which could have killed this new engineering firm. In the early hours of Friday 16 June 1843 a fire was spotted in Tuxfords' works. Workers and their wives living nearby rushed out to

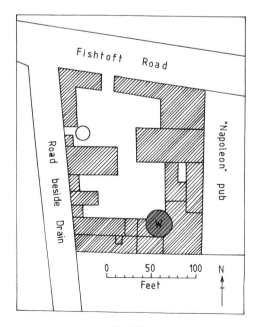

Fig 171
Layout of buildings on Tuxford's original site, based on plan with 1859 conveyance of extension site.

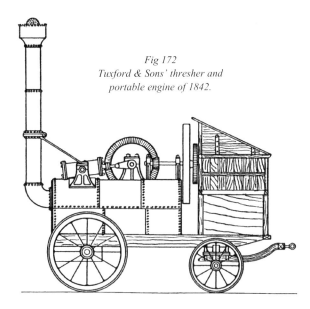

*Fig 172
Tuxford & Sons' thresher and portable engine of 1842.*

tackle the blaze, whilst others ran to Boston for the fire brigade. The fire was raging on the upper floor of the building containing the wooden casting patterns and by the time the fire engines arrived the roof had already fallen in. Efforts then turned to saving adjoining buildings, particularly the eight-sail windmill. Men turned the cap so that the sails were away from the inferno, and the mill and most of the site was saved. The firm had been working till 11 pm the previous night on some large castings and, though the cause of the fire was never found, it was assumed to be accidental. The damage was estimated at over £3,000, of which only part was insured. The loss included all of Tuxfords' patterns as well as the flour milling part of the site, though not the windmill itself.[15]

Much of the site had been saved, but the chaos could have severely interrupted business had not help come from a surprising quarter. Only a week after the fire Tuxford and Sons announced that 'through the kindness of Mr Caleb Taylor, who has allowed them the use of his Foundry, they are enabled to carry on the Founding Business without interruption until their premises are restored'.[16] Two weeks later they reassured their customers that as they kept a large stock of castings in hand they should be able to meet all orders as quickly as they had before. They were replacing their valuable stock of models and declared that they would not charge their customers for new models and patterns made to replace those destroyed by the recent fire. This latter notice was headed 'Boston and Skirbeck Iron Works', which is the first known use of the name.[17]

*Fig 173
Tuxfords' pile-driving machinery, shown in the western part of the works with timber stored for seasoning.*

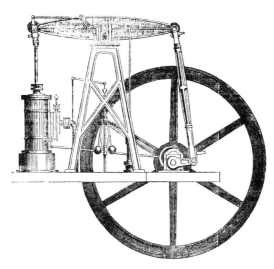

Fig 174
A small beam engine by Tuxfords
(from Illustrated London News, 22 December 1855).

Caleb Taylor's premises were at the opposite end of Boston, near the Grand Sluice, though their exact site is not known. To allow Tuxfords to use his premises must have caused problems for Taylor, so why did he do it? There is a clue in that James Edward Tuxford, the pawnbroker brother of W W Tuxford, had been listed as an iron and brass founder in 1826 with premises at the Grand Sluice.[18] J E Tuxford was not there in 1830 when Caleb Taylor was listed for the first time as an ironfounder in the same area. The possibility that Taylor was the tenant of J E Tuxford's premises is supported by his decision in 1843 to let W W Tuxford have the use of his premises at only a few days' notice.[19] The Tuxford family owned riverside marshland on the west side of Witham Place, directly opposite the junction with Union Place, so J E Tuxford's 1826 foundry might have been on the triangular site on the opposite side of the road where the Laughton/Blue Coat School was later built. We do not know when Tuxford & Sons returned to their Skirbeck premises, but it was probably within a year.

During the 1840s and 50s Tuxford and Sons greatly increased the range of their products and pressure on their small site must have become intense. In 1856 it was said that, as well as portable steam engines and threshing machines, they also made 'patent slip-ways for ship-yards, iron bridges, fixed steam-engines, wind, water and steam flour-mills, draining-engines, sluice works, elaborate and powerful steam pile-driving machinery etc.' *(Fig 173)*.[20] They made beam-engines *(Fig 174)* and in 1854 installed one at Timberland pumping station, which survived until 1926.[21]

In the 1851 census two partners both indicated that the firm employed 80 men and boys, but the 1856 *Directory* said the firm had 'several hundred hands', suggesting an indefinite, but probably large increase. In 1870 it was said to have 'ten score' (ie 200).[22]

William Wedd Tuxford's original windmill had been built in the countryside, with few houses nearby. As the business developed into an ironworks, some workers could walk the half mile from Boston, but for others a small number of houses arose closer to the works. There were some in Skirbeck Road to the west of Mount Bridge, others along the east bank of the Maud Foster Drain, and in Church Road were Maddison's Row and Napoleon Terrace. The public house built next to the works was also called the Napoleon, reflecting Tuxford's revolutionary political principles, in contrast to the Wellington pub about a mile away named after the Tory Prime Minister. For a time the Tuxford family had an attactive semi-detached house in Skirbeck Road, and Joseph Tuxford lived in the elegant Skirbeck Terrace on the bank of the Maud Foster Drain about half a mile north of the works.

In 1850 Tuxfords designed a new type of portable engine, known as a 'housed' engine, because most of the moving parts of the engine were enclosed in a cast iron housing at one end. The firebox was at the same end as the chimney.[23] In 1855 one of these engines was exhibited at the Paris Exhibition, where it drove machinery. That engine can still be seen in the Conservatoire National des Arts et Metiers in Paris *(Fig 175)*.[24]

Fig 175
Tuxford portable engine of 1855; still displayed in a Paris museum, though without its fold-down chimney (photograph 1989).

Fig 176
Tuxford engine of 1857-60 with Boydell's design of wheels, a valiant attempt at a self-moving engine.

Fig 177
Tuxford traction engine of 1861, now in a museum in Varmland, Sweden.

About the same time James Boydell invented a new design of traction engine and in 1857 Tuxfords were one of the firms that tried to make it a practical success. These three-wheel vehicles were built around Tuxfords' twin cylinder 'steeple' enclosed engine and return flue boiler portable engines *(Fig 176)*.[25] They sent several such engines to Cuba to work on sugar plantations, but in the end Boydell's system was not a success. In 1861 Tuxfords made a traction engine of a slightly different design and one was sold to a Swedish company that year. The engine in Sweden was soon converted to stationary usage with a belt drive from the flywheel and it still survives in the Hagfors Industri-och Jarnvagsmuseum in Motjarnshyttan,

Fig 178
1857 illustration of Boston and Skirbeck Ironworks; the original works on the left and the extension on the right.

Varmland, Sweden *(Fig 177)*.²⁶ The French, German and Russian governments all bought Tuxfords' engines as specimens for the instruction of their own engineering firms.²⁷ British-made portable engines were being exported to Australia in the 1850s and in 1859 Tuxfords won 'every prize' at the South Australia Agricultural Show. By 1860 Tuxfords had a branch in Adelaide, which also sold agricultural products of other British manufacturers. The next year they were exhibiting threshing sets at Malmo in Sweden through the firm of Jansson of Goteborg.²⁸

Fig 180
Woodworkers' shop; note no windows in the west wall, on the right overlooking the Rectory grounds (photograph 1966).

The Boston and Skirbeck Ironworks

Tuxfords needed space to expand and next to their works was the rest of the original Church Close, but while the Rev William Roy lived there was no chance of him letting them rent or buy it. Not only did he want to keep smoke and noise as far from his rectory as possible, but there were strong political differences between the radical ironmaster and the high Tory churchman. It is said that on one occasion when Mr Roy delivered a sermon against radicals, William Wedd Tuxford led his workers to 'ran-tan' his neighbour in his rectory, men and women blowing horns and drumming on pans to make 'rough music' to show their disapproval of what the Rector had said!²⁹ Mr Roy died on 2 October 1852 and his widow proved more amenable. Tuxford & Sons had already built a large new forge on the Church Close by 1857 and they purchased the field on 4 May 1859 for £1,000.³⁰

The conveyance from Mrs Roy to Tuxford and Sons imposed several restrictions on what could be done on the land. It specified how high the present and future buildings could be, that the firm could not build any further new furnaces on the land, and that they could not extend the forges any closer to the Rectory than the buildings on the original 1822 site. No workers' houses could be built on the land, and there could be no windows in the east wall of the long building nearest to the Rectory *(Fig 180)*. In addition, Mrs Roy expected them to stop smoke pollution - fires in the woodworkers' shop could only burn coke or wood - and 'as far as practicable' Tuxford and Sons should 'abate' smoke and other nuisances from arising from the works.

Plans with the 1859 conveyance included an outline of the buildings on the original 1822 site grouped around

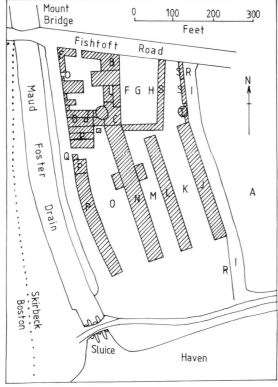

Fig 179
Plan of the 1859 extension to Tuxford's works, with buildings.

A: Skirbeck Rectory grounds
BCDE: Tuxford & Sons Ironworks
FGH: Napoleon Tavern and Wiseman's and Gibbs' houses
I: Private road to bungalow 30 feet wide
J: Woodworkers' shop
K: Yard
L: Proposed painters' shop
M: Yard
N: Turning, fitting and erecting shops
O: Yard
PPP: Smiths' and boilermakers' shop and forges
Q: Roadway by the drainside
RR: Ten feet high close fence
SSS: Sheds, stabling, etc
T: Gasometer for shop

Fig 181
Part of the Smiths' and Boilermakers' shops (west wall); this building survived until 2004 (photograph 1964).

a central yard, and showed the buildings already built or planned on the extension site *(Fig 179)*. A drawing of the works in 1857 *(Fig 178)* also shows the buildings of the old site (to the left) and extension (to the right). To the west of the works was the navigable Maud Foster Drain, which connected via the Frith Drain to the river Witham at Anton's Gowt and was used by barges bringing raw materials to the works from other parts of England. A crane on the wharf unloaded coal, coke, pig-iron, bar-iron, etc, and a little tramway carried them into and about the works.

The office and windmill were on WW Tuxford's original site, but by the late 1850s most of the work was done on the extension site in four long buildings parallel to each other. Next to the Maud Foster Drain were the smiths' and boilermakers' shop and forges. *(Figs 181/182)*. The smithy had eight iron chimneys with four hearths to each chimney, so that 64 men could work on 32 forges down the centre of the building.

On the east side of the site was the woodworkers' shop, which made the timber parts of threshing machines and other products *(Figs 180, 183)*. Between the two was the turning, fitting and erecting shop, where the machines were assembled, and parallel to it the painting shop, where they were finished before despatch. Around the yard were piles of iron and stacks of timber seasoning for up to five years.

As Boston gasworks was on the far side of town and the quality of pipes through the streets was not always good, Tuxfords' works had its own gasholder where a supply could be built up and used when needed without the risk of running out. In the north-east corner of the site were stables, for many products would be pulled by horse either to local customers or, more often, to Boston goods station, a mile away in London Road, for despatch by the Great Northern Railway. Until December 1884 the port of Boston could not take large ships, so it is unlikely that many of Tuxfords' products were sent out by sea from this port.

A visitor to the works in 1870 started his report by describing the office: 'The draftsmen were busy with pencil and compasses in a long upper room, marking-out the line for the busy colony of ten-score workers in wood and iron below.' The room was full of models that had been made to solve problems, including 'a traction engine with an endless railway attached' *[the Boydell system; Fig 176]*. 'Pigs of iron are piled in the yard below, and workmen are breaking them up for the furnaces.' On the other side of the yard was the moulding shop. 'In this shop, puddlers with brawny sinews and "auctioneers" (which election bullies have not cared to meet twice) are bending over huge casting boxes, or treading in the clay for a girder mould, as if they were working in a wine vat.' Overhead were giant cranes. In an adjoining shed were 'some grand left-handed hitters among the quartets which gather round the anvils, or close up the rivets of the engine boilers' as well as machines to punch holes through half-inch sheets of iron. 'The coach house [paint shop] is across the yard, and there stand upwards of forty engines ready for going out, and some of them packed-up for Japan. Blue was once the body colour, but of late years the taste of customers has run in favour of green.'[31]

Passing into the woodworking shop *(Figs 180, 183)*, seven threshing machines were receiving their last touches and 'patterns of wheels hang on the wall like shields'. There was 'a mysterious model gallery

Fig 182
The east wall, from within the yard, of the Smiths' and Boilermakers' Shops (photograph 1966).

Fig 183
Woodworking Shop, west side facing into yard. Note alternate spacing of original doors and windows, some later blocked up (photograph 1966).

running along the centre of the roof' where the wooden models were now kept. 'All the wood is seasoned for five years, under rain and sunshine in the yard. The elm and the ash are nearly all from the fens, and have 33 per cent more gravity in that rich clay loam than when grown on lighter soils. Revesby and Kirkstead have furnished many a stately oak, and there was a memorable purchase at Pinchbeck of three oaks growing from one stool, which fell before the wind in a night.'[32]

The cost of buying the site extension in 1859 and putting up new buildings meant that Tuxford and Sons had to re-arrange their finances. The £2,000 borrowed in 1837 had been partly paid and the rest 'otherwise satisfied', perhaps in part by putting up cast-iron gates at the home of the bank's local director living in Skirbeck. William Wedd Tuxford's original mortgage of 1822 had been reduced from £2,000 to £1,500 and in 1859 his sons agreed to share the debt with him. On 1 July 1859 the firm's account with the Stamford, Spalding and Boston Bank was in the red to the tune of £5,880.18s.0d, but at Tuxford's request the bank agreed to forego immediate payment on receiving further security. Tuxfords borrowed £2,000 from their cousin, George Parker Tuxford of London.

William Wedd Tuxford remained a partner in the firm for another seven years and only retired in 1866, when his four sons took over the firm completely. At that time he was again living above his baker's shop in the Market Place, having recently moved back. He died on 11 August 1871, aged 89, and was buried in the family vault in St Nicholas Churchyard.[33]

None of Tuxfords' business records survive, so we only have piece-meal information on the number and type of products they were making, but a notebook from the early 1860s gives some idea of their activities just after the move to their extended site. In 1860 they made 71 portable engines and 41 'agricultural machines' [probably threshing machines], a total of 112 items for £28,548.18s.0d, in addition to repair work and other outside jobs. In 1861 the notebook listed three 'locomotives' [meaning traction engines], 54 portables, 2 horizontal engines and 24 'agricultural machines'.

The following year they apparently made just one road locomotive, 40 portables, 2 horizontal engines, 25 threshing machines and four elevators.[34] They ceased making their 'housed' portable engines in 1863 and their later portables were more conventional.[35]

The Heyday and Decline of the Firm

In 1870 their European markets included France, Hungary and especially Austria for engines, finishing machines and centrifugal pumps. Orders had also been sent to New Zealand, Peking (Beijing), Shanghai, the Burra Burra mines in Australia, Cuba, Peru and California. To India the firm had sent sawing-machines for the teak forests of Burma, traction engines and trains of wagons to Calicut to bring coffee down from the plantations, and an engine to spin wire for the telegraph works of Bengal. Steam packing machinery for wool had been sent to the Government of Queensland, Australia and two fibre mills with hydraulic presses had gone to Loanga in Africa. During 1870 the firm was making a 30-foot waterwheel for Natal in South Africa and cast-iron parapets for the Thames Embankment in London.[36] The 1871 form of their traction engine is shown in Fig 170, that particular engine working on the Thorold estate near Grantham until 1914.[37]

In 1871 the firm gained a prize at the Royal Show for its detached steam-powered windlass for steam ploughing.[38] The 1872 *Lincolnshire Directory* listed a full and varied range of products for the firm and stated that it employed about 300 hands. This may be an exaggeration, as a book of 1870 said that they had 200 workers. Similar entries appeared in 1876 and 1882 directories, describing the firm as 'engineers, millwrights, smiths & engine boiler makers, iron & brass founders, makers of Tuxfords' patent straw elevators & land drainage & irrigating machinery'.[39] One of their more unusual products was a steam-driven fairground roundabout, a version of galloping horses, which they made in 1878 for a local firm.[40]

By the mid-1870s British agriculture was entering a period of decline and firms dependent on it were very vulnerable. By 1877 Tuxford and Sons' debt to the Stamford, Spalding and Boston Bank was over £6,000 and that debt was still on the books in 1911 when the bank itself was wound up. In 1879 Tuxford & Sons were laying off workers and the future did not look good. Joseph Tuxford died on 21 October 1878 and a year later the other three brothers dissolved the partnership. Wedd and William, now aged 73 and 59 respectively, formed a new partnership as millers and bakers and continued to use the windmill on the site to supply their Market Place shop.

Sixty-five year old Weston Tuxford carried on the engineering business alone, though still trading as Tuxford & Sons, and appointed William Towell (1837-1914) as a manager to assist him. But the firm was in the twilight of its days. Weston Tuxford, perhaps the real driving force behind the firm's creativity during most of its existence, had moved from 21 Market Place to bachelor rooms at the works office building and had a housekeeper to look after him. On the evening of 11 May 1885 Weston collapsed and was found by William Towell, who was working late. A doctor was called, but Weston had passed away before the doctor could reach him. The death of Weston was the real end of Tuxford and Sons, though it was another year before the contents of the site were sold off and a few more years before iron-founding finally ceased. The milling and bakery business of William and Wedd Tuxford was also affected, as their mill was part of the site taken over by the bank. By 1885 they had moved out of 21 Market Place and traded from rented premises in Shodfriars Hall, just south of the Market Place, but their business finally ended about 1891 when the mill was taken down.

The Boston and Skirbeck Ironworks passed into the hands of the Stamford, Spalding and Boston Bank and in July 1886 everything on the site, the remaining engines and other products and all the plant, machinery and tools, were sold in a three-day sale. It was a sign of the times that with agriculture in depression there was little demand for such items and the auction attracted only a moderate attendance.

In 1887 much of the site, equipment and materials were taken over by a new firm called Collett & Co, which rented the works from the Bank. By this date some of the buildings in the centre of the site extension had already been removed. Collett & Co declared their intention of recovering some of the lost trade and prestige of the concern and it is possible that the new firm was making engines and selling them under the Tuxford name, presumably using Tuxford designs and patterns, and probably also Tuxford workers.[41] Several of the Tuxford engines surviving in the twenty-first century seem to date from the Collett period. But after 'four years hard work and the loss of considerable capital', Collett & Co. withdrew, and on 19 November 1891 all the buildings were auctioned off. Those on the original site were pulled down and cleared away. The eight-sail windmill was bought by John Pocklington and the cap and sails transferred to a windmill at Heckington where it still survives. Rows of new houses were soon built on the site of Tuxfords' original works, but Mrs Roy's covenant of 1859 prevented houses being built on the extension site, so that remained in industrial use until 2004.

The woodworkers' shop survived until 1999 and part of the smiths' and boilermakers' shops and forges remained until 2004, but what appears to be the original office block on the corner next to Mount Bridge was still surviving in 2006, having been converted to a hairdresser's premises and other uses.

However rich Tuxford and Sons may have been in invention and ingenuity, the business did not apparently give the family much in the way of material prosperity. The firm's lack of profitability, and eventual demise, is partly due to the emphasis on innovation and development - the best way to a fortune was to find a successful design and then produce it in great numbers without much change. That was not Tuxfords' forte. The firm was happiest searching for solutions to new problems. It always did its own thing, did not follow designs developed by others, but always looked to find its own solutions and its engines had little in common with what became conventional simplicity.[42]

The firm served a world market and provided employment and training for many people in Boston and Skirbeck over several decades. Even its decline and closure in the 1880s was to bring benefits to the town as it coincided with the opening and development of Boston Dock, and firms that were arising to support the Dock and its fishing fleet were in need of skilled men. In 1919 George Palmer, the engineer of the Boston Deep Sea Fishing and Ice Company, was still proud to say that he had served his apprenticeship at Tuxfords' foundry.[43]

Fig 184
Tuxford 2 nhp portable steam engine, No 1283, built c1875-80, working a saw bench at Weeting Steam Rally, Norfolk, 2005

Notes

[1] Neil R Wright, 'Tuxford and Sons of Boston - a family business' in *Lincolnshire History and Archaeology*, Vol.38 (2003), p.61.

[2] *Stamford Mercury (LRSM)*, 24 September 1813, p.3, col.3; G S Bagley, *Boston Its Story and People* (1986), p.225; M Hanson and J Waterfield, *Boston Windmills* (1995), p.22.

[3] *LRSM*, 4 August 1837, p.3, col.1.

[4] Title deeds of Tuxfords' extension site, seen on 14 January 1999 in the Nottingham offices of the solicitors to Fogarty Ltd., through the good offices of the Managing Director of Fogarty Ltd.

[5] Op. cit.

[6] The Druid (H H Dixon), *Saddle and Sirloin or English Farm and Sporting Worthies (Part North)*, London (1870), pp.479-80.

[7] *LRSM*, 18 August 1837; N Leverett and M J Elsden, *Aspects of Spalding 1790-1930* (1986), p.34; the bridge collapsed 26 January 1845.

[8] Ronald H Clark, *Steam Engine Builders of Lincolnshire* (1955 reprinted 1998), p.113; W White, *Directory of Lincolnshire*, 1st edition (1842), p.164.

[9] *Boston Guardian*, 16 May 1885, p.5.

[10] *Illustrated London News*, 22 December 1855, Vol.27, p.726.

[11] *Agricultural Gazette*, 3 October 1874, p.1258; *Boston Guardian*, 25 October 1878.

[12] *LRSM*, 12 June 1840.

[13] *Illustrated London News*, 22 December 1855, Vol.27, p.726; Ronald H Clark, *Steam Engine Builders of Lincolnshire* (1955 reprinted 1998), pp.114-15.

[14] Sir William Tritton, 'The Origin of the Threshing Machine', *Lincolnshire Magazine 2*, (1934-36), p.8; *LRSM*, 3 March, 28 April and 23 June 1843; Ronald H Clark, *Steam Engine Builders of Lincolnshire* (1955 reprinted 1998), p.115.

[15] *LRSM*, 23 June 1843, p.3, col.3.

[16] *LRSM*, 23 June 1843, p.3, col.7.

[17] *LRSM*, 14 July 1843.

[18] W White, *Directory of Lincolnshire* (1826), pp.81, 83, 86.

[19] Pigot & Co., *Directory of Lincolnshire* (1830), p.512; LRSM, 23 June 1843, p.2, col.7.

[20] P. Thompson, *History and Antiquities of Boston* (1856 reprinted 1997), p348.

[21] Ronald H Clark, *Steam Engine Builders of Lincolnshire* (1955 reprinted 1998), p.125.

[22] The Druid (H H Dixon), *Saddle and Sirloin or English Farm and Sporting Worthies (Part North)*, London (1870), p.482.

[23] Ronald H Clark, *Steam Engine Builders of Lincolnshire* (1955 reprinted 1998), pp.114-15.

[24] P Thompson, *History and Antiquities of Boston* (1856 reprinted 1997), p348; Neil R Wright, 'Tuxford and Sons of Boston - a family business' in *Lincolnshire History and Archaeology* Vol.38 (2003), p.61.

[25] Richard Brooks, *Lincolnshire Engines Worldwide*, p.4.

[26] Richard Brooks and Martin Longdon, *Lincolnshire Built Engines*, p.30.

[27] Neil R Wright, 'Tuxford and Sons of Boston - a family business' in *Lincolnshire History and Archaeology* Vol.38 (2003), p.53.

[28] Richard Brooks, *Lincolnshire Engines Worldwide*, pp.12, 58, 62.

[29] Rimmington Wells, in a lecture reported in *Boston Guardian*, 8 October 1941.

[30] Neil R Wright, 'Tuxford and Sons of Boston - a family business' in *Lincolnshire History and Archaeology* Vol.38 (2003), p.54.

[31] The Druid (H H Dixon), *Saddle and Sirloin or English Farm and Sporting Worthies (Part North)*, London (1870), p.481-84.

[32] *Ibid.*

[33] *LRSM*, 18 August 1871, p.4, col.4.

[34] Tuxford notebook quoted in R A Clark, *A Traction Engine Miscellany* (1975), pp.84-85.

[35] Ronald H Clark, *Steam Engine Builders of Lincolnshire* (1955 reprinted 1998), p.119.

[36] The Druid (H H Dixon), *Saddle and Sirloin or English Farm and Sporting Worthies (Part North)*, London (1870), p.481-84.

[37] Ronald H Clark, *Steam Engine Builders of Lincolnshire* (1955 reprinted 1998), p.129.

[38] *Agricultural Gazette*, 3 October 1874, p.1258.

[39] W White, *Directory of Lincolnshire* (3rd edition 1872), p.819; E R Kelly, *Post Office Directory of Lincolnshire* (5th edition 1876), p.61; W White, *Directory of Lincolnshire* (4th edition 1882), pp.161, 162.

[40] Ronald H Clark, *Steam Engine Builders of Lincolnshire* (1955 reprinted 1998), pp.132, 133.

[41] Ronald H Clark, *Steam Engine Builders of Lincolnshire* (1955 reprinted 1998), p.24; Bagley, *Boston Its Story and People* (1986), p.226; *Kelly's Directory of Lincolnshire* (2nd edition 1889), p.71.

[42] Richard Brooks, *Lincolnshire Engines Worldwide*, p.4.

[43] *Boston Guardian*, 30 August 1919.

LINCOLNSHIRE IMPLEMENT MAKERS: A SELECTED LIST

Over the past two centuries, hundreds of Lincolnshire men have turned their hands to making machines and implements for the county's prime industry - farming.

Most of these individuals worked in their local town or village with only limited success and both their names and their products are now largely forgotten. However, a surprisingly large number were making machinery for at least two decades. The list that follows sets out the names of these, the relatively successful and long-lasting firms.

The selection has been based largely on the trade directories for the county covering the period from the early nineteenth century up to the Second World War. The first widely available countywide directory was published by Pigot in 1822, the last by Kelly in 1937. Hence firms whose operating dates are given as commencing in 1822 may well have been in existence several years before then, and, for a similar reason, many of those indicated as finishing in 1937 continued in business after that date.

In some instances there are gaps of four or five years between the directories available and thus the dates given below only indicate approximate periods of operation of firms. It should also be noted that information given by trade directories is not always reliable.

Several firms passed through more than one generation of ownership. For example, as can be seen in the list, the Barlows, implement makers of Kirkby la Thorpe, traded under five different names between 1882 and 1937.

Name of Firm	Dates	Location
Alford, John,	1820-1841	Holbeach
Archer, Henry	1849-1876	Barrowby
Ashley, Thomas & Sons	1856-1909	Louth, Cannon Street, 1856-76; Aswell Iron Works, 1876-1909
Ashton, William & Son	1868-1905	Horncastle, Hopton Iron Works, Boston Road
Barlow, John	1882-1909	Kirkby la Thorpe
John Barlow & Son	1913-1919	
Barlow & Twells	1922-1926	
Cecil William Barlow	1926-1930	
Walter Barlow & Son	1930-1937	
Barlow, Joseph,	1876-1900	Coningsby
William Barlow	1900-1905	
Barrett, Lucy	1822-1835	Lincoln, St Mary's
Barrett, Arthur Y & Co	1842-1855	Horncastle, Union Foundry, Union St
Barrett, Joseph,	1822-1835	Boston, West St, 1822-51;
Elizabeth Barrett	1841-1851	
Edwin Barrett	1856-	Boston, Horncastle Rd, 1856
Barton, William	1861-1876	Boston, Strait Bargate
Beacock, Matthew	1826-1841	Winterton (See Fletcher of Winterton)
Bell, John Andrew	1909-1937	Stainfield and Bardney
Blackstone & Co		
Smith & Co	1844-1851	Stamford, Rutland Terrace Iron Works, St Peter's Street, 1844-1887
Smith & Ashby	1851-1859	
T W Ashby & Co	1859-1864	
Ashby & Jeffery	1864-1866	
Ashby, Jeffery & Luke	1866-1876	
G E Jeffery & Co	1876-1877	
Jeffery & Blackstone	1877-1882	
Blackstone & Co	1882-1889	Stamford, Ryhall Road, 1887-1994
Blackstone & Co Ltd	1889-1936	
Lister Blackstone & Co Ltd	1936-1965	
Mirrlees-Blackstone Ltd	1965-	
Bradbury, Thomas	1849-1890	Gainsborough
Richard Bradbury	1890-1893	
Bradshaw, John	1896-1933	Sturton by Stow
John Bradshaw & Sons	1937-	

LINCOLNSHIRE AGRICULTURAL IMPLEMENT MAKERS

Bradshaw, William	1856-1882	Boston, Shodfriars Lane, 1856-1861; 12 Sibsey Lane, 1861-82
Brooke, Thomas,	1856-1896	Market Rasen, Linwood Rd
Thomas Brooke & Son	1900-1919	
William Brooke	1922-	
Burr, John	1849-1876	Long Sutton, Chapel Bridge
Caborn, John	1849-1872	Denton
Richard S Caborn	1876-1882	
Clarke, James	1856-1868	Lincoln, 46 Waterside South; Lincoln Crank and Forge Works from 1868
Clarke, Edward	1868-1872	
Clayton & Shuttleworth	1842-1936	Lincoln, Stamp End Works
Coldron, Frederick	1919-1937	Wellingore
Cooke, John	1841-1887	Eagle, 1841-55; Lincoln, 20 Monks Road, 1855-70;
John Cooke & Sons	1887-1938	Lindum Plough Works, Monks Road; 1870-1936
Coulson, John	1868-1872	Stamford, St Paul St, 1868-72;
Coulson & Wear	1872-1882	Rock Ironworks, Scotgate, 1872-82
Coultas, James	1852-1921	Grantham, Wharf Road, 1852-62; Perseverance Works, Station Road, 1862-1955
James Coultas Ltd	1921-1955	
Crawford, Robert Henry	1926-present	Frithville
Drury, John	1882-1909	Laceby, Cross Roads
John Drury & Son	1913-1926	
Duckering, Richard & Burton	1845-1852	Lincoln, Waterside South, 1845-58;
Richard Duckering	1852-1870	Lincoln, Waterside North, 1858-1965
Charles Duckering	1870-1917	
Richard Duckering & Co	1917-1962	
Edgley, William	1882-1913	Sutton Bridge
Edlington, John B & Thomas	1865-1893	Gainsborough, Phoenix Iron Works, Lea Rd
J B Edlington	1893-1905	
J B Edlington & Co Ltd	1905-present	
Elvin, James	1842-1872	Sleaford, North St
Farmer, Thomas	1841-1856	Gainsborough, Trent Bank
Thomas Farmer & Son	1861-1880	
Farmer, Robey, Brown & Co	1880-1884	
Farmer, Robey, Clark & Co	1889-	
Fenton, Joseph Bentley	1872-1922	Great Hale, 1872-1893; Sleaford 1893-1937
Fenton & Townsend Ltd	1937-	
Fenton, William Bentley,	1889-1922	Eagle, Acme Iron Works
Fletcher: (previously Matthew Beacock, 1826-40);		
Beacock & Fletcher	1840-1852	Winterton, Holly House, Wintringham Road, 1826-52;
J Fletcher	1852-1877	Winterton, Newport Ironworks, 1852-present
T & J Fletcher	1877-1885	
T Fletcher	1885-1898	
T Fletcher & Son	1898-1925	
H J Fletcher	1925-1935	
T & J Fletcher	1935-1992	
T & J Fletcher Ltd	1992-present	
Fletcher & Cook	1841-1868	Epworth Ferry/Owston Ferry
James Fletcher	1868-1872	
Fletcher, William	1876-1896	Aslackby
Foley Brothers	1893-1919	Bourne, North St & Meadowgate
Ernest Alfred Foley	1922-1930	
Foster, William	1856-1877	Lincoln, Wellington Foundry
W Foster & Co Ltd	1877-1960	
Fox, George Maples	1861-1876	Lincoln, 206/7 High St, 1861-76;
Fox & Sykes	1882-	Lincoln, Free School Lane, 1882
Gibson, Samuel	1861-1882	Barton, Mabel Foundry, Brigg Rd

Gibson, Thomas	1856-1893	Stamford, 22 Broad St
Grant, Isle	1856-1889	Binbrook
Grassam, Seth	1849-1872	Spalding, Crescent Ironworks, Sheep Market
Gratton, David Thomas	1905-1913	New Leake, 1905-1919;
D T Gratton & Sons	1913-1937	Boston, 13 Wide Bargate, 1922-37
Graves, Thomas	1849-1865	Old Bolingbroke
James Graves	1868-1893	
Gray, Charles	1889-1896	Stamford, 1 St Mary's Hill, 1889-96;
Charles Gray & Co	1900-1905	Stamford, Castle Hill Foundry, 1900-05
Greenfield, Elijah	1861-1905	Market Deeping, Halfleet
Grounsell, John	1826-1851	Horncastle, Foundry St
James Grounsell	1856-1856	
Grounsell, Frederick	1828-1856	Louth, Westgate, 1828-1889;
H Grounsell	1861-1872	
Turner & Grounsell	1876-1889	
F Grounsell & Son	1893-1937	Louth, Northgate Ironworks, 1893-1937
Haith, William	1868-1909	Keelby
Mrs C Haith	1913-1926	
Harper, George	1870-1882	Grimsby, Eleanor St
Alfred Drewery	1882-1892	
Drewery & Harper	1892-1900	
Harper & Co	1900-1909	
Harper, Phillips & Co Ltd	1913-1937	
Harris (& White)	1856-1865	Sleaford, Old St;
Harris, William Robert	1868-1889	Sleaford, Boston Road
Harris, Smith & Co	1893-1933	
Harrison, Thomas	1861-1882	Lincoln, 5 Burton Road
Harrison, Teague & Birch	1874-1904	Lincoln, Brayford Wharf, 1874-1922;
Harrison & Co	1904-1938	North Hykeham 1905-2006
Leys Malleable Castings Ltd	1938-1980	
George Fischer (Lincoln) Ltd	1980-2004	
Lincoln Castings	2004-2006	
Hart, James & Wm. Berridge	1822-1825	Brigg, Bridge St; Ancholme Iron Works, 1824-72;
James Hart	1825-1849	also Caistor, Grimsby Road, 1841-68
James Hart & Son	1849-1872	
William Hart	1872	
Hart, Walter	1930-1937	Anwick
Haslam, William	1861-1919	Holbeach, Boston Rd Iron Works
Hayes, William	1893-1913	Holbeach, 1893-1896; Sutton Bridge 1896-1913
Hempstead, C H & Co	1900-1926	Sleaford, Cane St, 1900-13; Grantham Road, 1913-26
Hempsted, Robert;		
Hempsted & Felton	1868	Grantham, George St; Phoenix Foundry, 1868-
Hempsted & Co,	1876-1885	
Grantham Crank & Iron Co	1885-1905	
Grantham, Boiler & Crank Co	1905-	
Hett, Charles Louis	1872-1895	Brigg, Ancholme Foundry
Hodgson, John	1872-1905	Louth Newmarket/Upgate, 1872-1922
Hodgson & Son	1913-1919	Louth, Albion Place, Eastgate, 1922-30;
J Frederick Hodgson	1922-1930	also Spilsby, Reynard St, 1913-
Holt, J	1868-1889	Donington
George Holt	1893-1909	
Hornsby & Seaman	1815-1828	Grantham, Spittlegate Ironworks;
Richard Hornsby	1828-1851	
Richard Hornsby & Son	1851-1879	
Richard Hornsby & Sons Ltd	1879-1918	
Ruston & Hornsby	1918-1961	Grantham and Lincoln (see Ruston)

Howden, William & Son	1803-1860	Boston, Stell's La/Phoenix Foundry
Hundleby, John	1893-1933	New Leake
Hunter, George	1849-1889	Ulceby
Hunter, William	1849-1876	Binbrook
Ingledew, Edward	1822-1849	Gainsborough, Bridge St
Jennings, Chas Dickinson	1856-1876	Spalding, New Road; Later in Bridge St, then Winsover Road
Johnson Bros	1905-1937	Bourne, Market Place & South St; Spalding, 1 Churchgate, 1922-37
King, Arthur S	1926-1937	Market Rasen, 24 & 26 Oxford St
Kittmer, Benjamin	1856-1889	Fulstow
Kynman, Thomas	1822-1861	Caistor, Grimsby Road
Laming, George	1893-1896	Owston Ferry
William Laming	1900-1919	
John W Laming	1922-1937	
Leak, TT	1822-1865	Boston, Stell's Lane, Skirbeck, 1822-65;
John C Leak	1868	Boston, 67 High Street, 1868
Lewin, Joseph	1861-1876	Ropsley
Malkinson, Henry	1822-1842	Boston, West St
Marfleet, Joseph	1826-1861	Wainfleet All Saints
Marris, Charles	1876-1909	Kirton in Lindsey, Albion Foundry
Marris Brothers & Beverley	1913-1937	
Marshall, William	1842-1857	Gainsborough, Britannia Iron Works
Marshall, William & Sons	1857-1862	
Marshall, Sons & Co	1862-1969	
Marshall-Fowler	1969-1975	
Aveling-Marshall	1975-1979	
Track-Marshall	1879-1985	
Martin's Cultivator Company	1902-1937	Stamford, Ryhall Rd. (Harrison Patents previously)
Mason, Robert	1849-1865	Alford, Market Place
Mrs Eliza Mason	1868-	
Mrs E Mason & Sons	1872-1889	
Mason & Son	1893-	
Mettam, Thomas	1896-1937	Sleaford, Wharf House, Carre St
Morley, John	1893-1896	Grantham, Wharf Road, 1893-1900;
Edward Morley	1900-1932	Grantham, Harlaxton Road, 1900-32
Morton, Joseph	1841-1882	Louth, 72 Eastgate, 1889-1913;
Morton & Son	1889-1909	
Morton, Son & Lock	1913-1937	Louth, Queen St, 1913-37
Nowell, William	1893-1919	Bourne, Abbey Rd
Thomas William Nowell	1922-1929	
Pape, Clarkson	1822-1841	Brigg, Silversides
Peacock & Binnington	1894-present	Brigg, The Old Foundry (& Hull)
Pearson, Matthew	1900-1926	Lincoln, Boultham Iron Works
Penistan, Michael & Co	1849-1870	Lincoln, St Rumbold's Lane
Penney & Co Ltd	1872-1910	Lincoln, 6 Broadgate & City Iron & Wire Works, 1872-1961;
Penney & Porter	1910-1968	Lincoln, Outer Circle Road, 1961-68
Pickwell, Thomas	1868-1905	West Torrington
Porter, J T B	1868-1876	Lincoln, Gowts Bridge Works, 1868-1910;
Porter & Co	1882-1913	
Penney & Porter	1922-1937	Lincoln, Broadgate, and Saxilby, 1910-37
Rainforth, William	1868-1889	Lincoln, Britannia Iron Works
Rainforth, W & Sons	1893-1933	
Ranby, Edmund	1856-1882	Tetford, 1856; Bilsby, 1868-1900
Joseph Ranby	1893-1900	
Rawlinson, T & J	1872-1900	Caistor, Quarry Terrace
Reed, Walter T	1905	Epworth
Reed & Sons	1905-1922	

Revill, Charles	1835-1889	Lincoln, 233/195 High St, Corn Hill, 1835-89;	
Frank Clarke Revill	1889-1913	Swanpool Court, 1889-1913	
Robey, Robert	1854-1861		
Robey & Co	1861-1969	Lincoln, Perseverance Works, Canwick Road, 1861-1969	
Robinson, George W	1828-1866	Barton, Market Lane, 1828-1868	
Robinson & Neave	1868		
Neave	1872-1876	Barton, George St, 1872-76	
Rundle, John H	1913-present	New Bolingbroke	
Rushby, Charles Thomas	1893-1937	Ruskington, High St	
Russell, W	1849-1876	Kirkby cum Osgodby	
Ruston			
Proctor & Burton	1840-1857	Lincoln, 48 Waterside South	
Ruston, Proctor & Co	1857-1918	Lincoln, Sheaf Iron Works, Waterside North	
Ruston & Hornsby Ltd	1918-1961		
Sanderson, John	1822-1882	Louth, Newbridge Ironworks	
John Sanderson & Son	1889-1905		
Sargeant, James William	1919-1922	Frithville	
Jesse Sargeant	1926-1937		
Savage, Thomas	1842-1868	Crowle	
Scaman, Matthew	1868-1882	Grimsby, South St, St Mary's Gate	
Mrs Maria Scaman	1893-1905		
Scaman, Newcombe Matthew	1872-1893	Horncastle, 45 Foundry St	
Scoffield, Levi	1872-1919	Sudbrook, Grantham	
William Scoffield	1922-1937		
Scunthorpe Foundry Co. Ltd	1919-1937	Scunthorpe, Dawes Lane	
Shores, J & Co	1896-1913	Owston Ferry, Trent Ironworks	
Simpson, Ann & Thomas	1849-1855	Lincoln, 233 High St, 1861-68;	
Thomas Simpson & Co	1861-1893	Lincoln, 292 High St, 1876-93	
Sinclair, John	1876-	Horbling, 1876	
William Sinclair	1882-1889	Billingborough & New York, 1882-1900;	
William Sinclair & Son	1893-1926	New York, 1900-26	
Smith, Edward H	1863-1913	Brigg, 2 Market Place	
E H Smith & Sons	1922-1926		
Spight, Isaac	1849-1892	Brigg, Victoria Foundry, 1849-68; Bigby St, 1876-96	
I Spight & Son	1896		
Spray, Samuel	1835-1855	Gainsborough, Mart Yard, 1861-72;	
Mrs A Spray	1861-1872		
Henry Spray	1876	Gainsborough, Lord St, 1876	
Stanton, Charles	1900-1937	Spalding, 21 Winsover Rd	
Stephenson, Charles Tooley	1872-1876	Boston, Mount Bridge Ironworks, Skirbeck, 1872-76;	
Thomas T Stephenson & Son	1876-1900	Boston, Vauxhall Works, Tower Rd, Skirbeck, 1876-1937	
H Stephenson	1905-1926		
Thomas Stephenson & Son	1930-1937		
Stephenson, William	1849-1882	Haxey	
Store, Robert	1876-1905	Alford, South End	
Taylor, Caleb	1822-1849	Boston, 1822-35; Alford, South End, 1849	
Thompson, Alexander	1861-1909	Keelby	
George M Thompson	1913-1919		
Thompson, George	1822-1865	Boston, Spilsby Road	
Tong, Edmund	1872-1882	Lincoln, Westgate Implement Works, 1872-1930;	
Mrs Edmund Tong	1889-1919		
George M Tong	1922-1937	Saxilby, 1933-37	
Tuxford, William	1824-1838	Boston, Skirbeck Ironworks	
William Tuxford & Sons	1838-1887		
Twidale, E	1919-1922	West Butterwick	
E Twidale & Sons	1926-1937		

LINCOLNSHIRE AGRICULTURAL IMPLEMENT MAKERS

Waite, George	1876-1905	Croft, 1876-1900; Wainfleet All Sts, St John St, 1893-1905
Frederick Waite	1913	
Ward, James & Co Ltd	1872-1933	Horncastle, North St, 1872-1933; 5 Bull Ring & Banks St, 1900-33
Warren, George Anthony	1913-1922	Horncastle, Wong
G A Warrren & Sons	1926-1937	
Watkinson, William	1835-1872	Louth, Newmarket Iron Works, 1835-68
William & George Watkinson	1876-1882	Louth, Cemetery Lane, 1876-82
Watson, Thomas	1876-1889	Spilsby, Alma Ironworks, Halton Rd
T Watson & Son	1893-1909	
Westoby, Alfred	1872-1896	South Kelsey
Westoby Bros	1900-	
White, Walter	1896-1919	Markby, 1896-1900; Hannah, 1909-1913;
W White & Son	1922-1937	Alford, Market Place, 1922-37
Whitton, William	1826-1868	Horncastle, Far St,1826; Foundry St, 1835-55; Spilsby Rd, 1861-68
Wilkinson, Wright & Co	1850-1860	Boston, Grand Sluice Iron Works
Henry Wright	1863-1876	
Winn, Edward	1922-1937	Stickney
Woodward, Joseph	1841-1861	Holbeach
Wright, John	1876-1893	Spalding, Halmer Cottage

Fig 185
Portable steam engine made by Thomas Harrison of Burton Road, Lincoln in 1863.

POSTSCRIPT

The editor would welcome additional information or comment about any of the implement makers and their products featured in this book.

In particular, further details are sought about the less well-known firms in the long list. For example, is it known exactly when the firms started or finished? Where were their workshops in the village or town? Are there photographs of buildings and workmen? Do some of their implements and machines survive? Are there newscuttings about the firms or their products? Are there any surviving family members? Have any notable firms been omitted from the list?

Please write to: The Editor **(Ken Redmore)** at the *Society for Lincolnshire History and Archaeology*, Jews' Court, 2-3 Steep Hill, Lincoln, LN2 1LS. The Society intends to publish more accounts of Lincolnshire agricultural engineers and implement makers in the future.

GLOSSARY

Page numbers refer to illustrations of the implement or machine

Barn machinery: the collective term used for machines that were usually installed and operated in the barn or other farmstead buildings. These could include threshing machine, chaff cutter, mill, kibbler, oil-seed cake crusher, root pulper/slicer, winnower.

Cambridge roll or roller: an implement for crushing clods of earth and firming the ground after ploughing or harrowing. Sets of closely packed cast iron discs, with projecting ridges around their circumferences, run on a horizontal axle. Widths vary. (pp.34, 74)

Chaff cutter: a machine for chopping straw into very short lengths (1-2cm) in the preparation of feed (which provides bulk but has little nutritional value) for animals, especially horses. Some machines operate on a guillotine principle but the more common type has a large spoked wheel (hand, steam, or engine-powered) incorporating two or more sharp radial blades which cut the straw. There is usually a linked mechanism which feeds the straw to the cutting position. (pp.9, 71)

Chavings: a product of the mechanical threshing process consisting mainly of the short thin stem ends to which cereal grains have been attached. The separated husks around the grain are known as chaff.

Clod crusher: a general name given to implements for breaking down large clods of earth after ploughing or cultivating. It is usually some type of heavy roller made of cast-iron rings or spikes. Heavy weights are often added to improve performance.

Combine harvester: a machine which reaps, threshes and separates cereal grain (or other seed-bearing crops) in a single operation in the field.

Cultivator: an implement for working the soil to a reasonable depth without turning it over like a plough; weeds, roots and large stones may be brought up. The usual design has strong forward-pointing sets of iron or steel tines. (Types: drag, grubber, scarifier, scuffler). (p.153)

Dray: a flat-bed or low-sided, four-wheeled vehicle drawn by horse or tractor. With good manoeuvrability it was widely used for the delivery of goods in towns. It was also known as a lorry, lurry, or trolley. (p.26)

Dresser, corn: a machine operated by hand or engine for cleaning cereal grain by a combination of winnowing and riddling, thus removing most weed seeds, chaff and stones. (p.13)

Drill: an implement for the controlled sowing of crop seed. Different models have evolved for different seeds; the common one for cereal grain has a number of evenly spaced coulters creating tiny furrows into which the seed is dropped from flexible tubes fed via feeding mechanisms from a long wooden box which holds the stock of seed corn (grain). (pp.41, 45, 152)

Elevator: a machine for raising loose hay, straw or sheaves from ground level to the highest point of a stack. It operates on a conveyor principle and has adjustable pitch to accommodate the growing stack; the bed of the conveyor track usually folds in two and is carried on road wheels so that the elevator can be towed along as part of a threshing set. (pp.13, 43, 60, 102, 124)

Furrow press (or land press): it consists of one or two rows of heavy, sharp-pointed rings and is often trailed behind a plough to firm the soil before a crop of wheat is sown. (p.128)

Hames: the frame, originally made of wood, but later of iron or brass, around the horse's collar to which ropes or chains of the harness are attached.

Harrow: a family of implements of varied design which work the soil surface to produce a fine tilth and cover seeds after sowing. Larger harrows, like light cultivators, are used to remove weeds; chain harrows improve the condition of grassland. (p.156)

Haymaking machinery:

Mower: closely related to the cereal grain harvest reaper, the usual model for cutting grass has a horizontal reciprocating broad-toothed knife running between metal 'fingers'. Motive power is transferred from two travelling wheels to the cutting apparatus. (pp.60, 68)

Horse Rake: a set of closely-mounted curved steel tines held in a frame between two large wheels skim the ground to pick up cut the hay. A lever can be used to raise the tines to release the collected hay. (pp.8, 14)

Swath Turner and Tedder: machines for turning over, spreading or moving rows (swaths) of cut hay to help it dry in sun and wind. Many designs have been produced. (p.14)

Horse-gear: a portable source of power for barn machinery or easily driven machines such as a straw elevator. Horses harnessed at the end of wooden shafts walk in a circle to rotate a toothed drive wheel which is geared to a horizontal 'tumbling' shaft at ground level. (The earlier and larger fixed horse-gin in a permanent building with overhead shafting using two or four horses is rare in Lincolnshire.) (pp.43, 115)

Horse hoe: a family of implements for clearing weeds between rows of growing crops, eg wheat, turnips, potatoes. The most basic model has a land wheel at the front to control the depth of the hoe in the soil and is guided by handles held by the operator who walks behind. (p.152)

Horse Power: a measure of the power output of an engine (superseded by the kilowatt). The nominal horse power (nhp)

of a steam engine was a rough indication of power based on cylinder size. Brake horse power (bhp) of an engine is based on actual measurement of output, usually at the flywheel.

Hummeller: a machine for removing the awns ('beard') from barley. The most common model consists of a wire gauze cylinder inclined at angle of approximately 45° through which the barley grains are flung by revolving beaters.

Iron types:

Cast-Iron: iron with a high carbon content which is relatively brittle and weak in tension. It cannot be forged or worked mechanically, but can be easily melted and cast in moulds.

Wrought Iron: a relatively pure form of iron which is strong in tension. It can be worked to shape easily by hammering (usually when red hot) and is resistant to corrosion.

Steel: mild steel is iron with a small proportion of carbon; other steels are alloys of iron with metallic elements such as manganese, chromium and nickel to make them tough, stainless, suitable as edged tools, etc.

Kibbler: a mill for crushing or cracking cereal grain (often oats), peas or beans for animal feed. The grinding surfaces may be a pair of stones as in a conventional flour mill, but some kibblers have adjustable cast iron plates or rollers.

Lurry or lorry: see dray.

Manure distributor: a variety of machine which distributes dry manure (bone dust, soot, guano) in a controlled fashion eg from hoppers via fluted rollers, paddles or plates. Also a type of drill which delivers liquid farmyard manure at a regular rate down pipes alongside cereal grain.

Mill: A means of grinding cereal grain for animal feed. The grinding surfaces may be pairs of stones or adjustable cast-iron plates or rollers. Sources of power on the farm are occasionally water or wind, but mills are usually powered by steam, oil or petrol engine. (pp.13, 20, 23, 49)

Moffrey: (contraction of the word 'hermaphrodite') a two-wheeled cart convertible to a four-wheeled wagon-type vehicle by addition of a fore-carriage, platform and shelving (open framework extensions). It was commonly used in the eastern counties to provide more capacity for bulky loads at harvest time. (p.24)

Mower: see haymaking machinery

Oil-seed cake crusher: a simple machine for breaking large slabs of consolidated rape, linseed or cotton seed oil husks after oil extraction into small pieces suitable for animal feed. It consists of one or more pairs of spiked rollers mounted in a cast iron frame which are turned by hand or engine power.

Plough parts:

Coulter: usually a sharp knife-shaped piece of iron bolted to the beam of the plough, positioned so that its point is immediately before the share, its function being to slice vertically into the soil.

Share: a triangular piece of iron with a sharp edge fixed to the front of the mouldboard so that it makes a horizontal cut beneath the furrow slice.

Mouldboard or breast: made in cast-iron or steel, it has a twisted or contoured surface designed to turn over the furrow slice of soil and lay it alongside the preceding one. Varied shapes are suited to different conditions and purposes, eg ploughing grassland, competitions.

Plough types:

Swing plough: with no wheels, the depth and width of furrow are controlled by the draught settings and the ploughman's skill.

Wheel plough: the depth and width of the furrow are controlled by the draught settings. (pp.25, 75)

Double plough: has two separate mouldboards and creates two furrows. Usually requires three or four horses. (p.33)

Double mouldboard/ridging plough: has a pair of mouldboards mounted side-by-side at opposing angles which together form raised beds or ridges for potatoes and root crops.

Turn-wrest plough: also has two mouldboards side-by-side at opposing angles, but they are brought into use alternately depending on the direction of travel of the plough and having the effect of leaving all the furrows turned to one side. (p.32)

Balance plough: has the same effect as the turn-wrest plough. Identical sets of plough parts are mounted in opposing directions on an oblique-angled frame. At the end of each furrow the plough is tipped about the balance point to bring the second set of plough parts into operation without having to turn the plough round. Most commonly used in steam cultivation.

Sub-soil plough: has no mouldboard and is designed to break up the sub-soil without turning the soil above.

Mole plough: creates a channel at some depth below the soil surface for the laying of field drains.

Potato lifting plough: a plough frame fitted with a set of iron fingers which run below the potatoes and lift them to the surface for hand or machine picking.

Potato spinner: a simple machine which is pulled down the ridges of potatoes when they are harvested. Spinning metal arms throw the potatoes out and they are picked up by the workers following behind. (p.153)

Potato sorter: a machine with a series of conveyors and riddles to separate soil from harvested potatoes. It also sorts potatoes according to size before bagging or bulk transport and enables diseased or damaged potatoes and stones to be separated by hand. (pp.64-65)

Ram pump: a simple and robust pump with no moving parts except valves. It uses the energy from a large amount of water falling a short distance in a sequence of pulses to pump a smaller volume up to a much greater height. (p.91)

Reaper: a machine with a horizontal reciprocating broad-toothed knife for cutting ripened cereal grain at harvest time. As it is cut, a large revolving reel with horizontal wooden slats push the standing grain stalks over. Later models (called binders) have moving canvas belts which carry the cut stalks to a mechanism for binding them into sheaves with wire or twine. (pp.58, 63, 68)

Reeing machine: a riddling machine for separating different types of seed. The riddles are made from wire mesh of various gauges.

Roller, land: a term usually reserved for heavy cast-iron roller with flat surface. See clod crusher and Cambridge rolls.

Root pulper/slicer: a machine for shredding, slicing or pulping turnips or mangolds for animal feed. Roots are fed into a hopper and sliced or shredded by knives on the surface of a roller or disc turned by hand or engine power. (pp.21, 58, 72)

Rulley: a low open-sided wagon (cf. dray).

Sack barrow: a sturdy wooden frame fitted with a pair of wheels and a large D-shaped iron bracket for manoeuvring full sacks of grain around the farmyard and granary.

Sack lifter: a simple machine operated by hand crank for raising full sacks of grain to shoulder height for carrying.

Steam cultivation: huge ploughs (and occasionally cultivators, harrows and drills) were drawn across fields by steel cables attached to steam powered traction engines.

Steam engines:

Stationary engine: a beam, grass-hopper or horizontal engine fixed to the barn floor for driving machinery using belts and shafts. They were only installed on the largest farms. (p.136)

Portable engine: steam engine and boiler mounted on wheels and moved around by horses attached to the front wheel axle. It drove machinery from belting off the fly-wheel. (pp.10, 11, 43, 79, 98, 135, 136, 149)

Traction engine: self-propelled steam engine, which is commonly used as a power source for ploughing, cultivation, haulage and driving stationary machinery (eg threshing, milling). (pp.122, 132, 137)

Steam Tractor: a lightweight traction engine developed as a source of power, mainly for road use. (p.122)

Straw trusser: one of the varieties of machines that were developed in the evolution of the modern baler. Straw (or hay) was compressed under weights or possibly by pistons under steam power and then tied with wire into compact bundles for economical storage and convenient transport. (p.50)

Threshing machine: the cereal grain is knocked out of the ears by rapidly revolving horizontal wooden beaters in a cylindrical drum. This simple principle applies to both early small hand-powered machines and the later fixed water or horse-powered models. The common travelling machine, powered for so long by steam traction engine, also contains mechanisms for separating straw, chavings and chaff, as well as for cleaning and grading the grain. (pp.125, 135)

Threshing set: most commonly a traction engine, threshing machine and straw elevator, which moved from farm to farm in the autumn and winter months. (pp.123, 133)

Water Turbine: a flow of water is used to drive an encased wheel revolving on a vertical axis which in turn provides power to machinery such as a corn (grain) mill or a pump. (pp.92, 93)

Whippletrees or swingletrees: they allow two or more horses to draw a load such as a plough efficiently. Slim horizontal sections of timber (trees) and sets of chains link the horses through the hames on their collars to the implement.

Winnowing machine: usually a small hand-operated machine for separating threshed cereal grain from husks and other chaff or lightweight material. It consists of a wooden box through which the material is passed; by turning a handle attached to a powerful fan, a strong draught blows lightweight impurities to one side while the heavier grain drops down. Later threshing machines incorporate a winnowing machine and were known at the time as combined threshers.

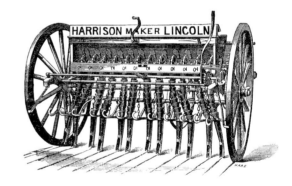

Fig 186
Corn drill, Thomas Harrison, Lincoln, 1868

Fig 187
Horse hoe, Isaac Spight, Brigg, 1868

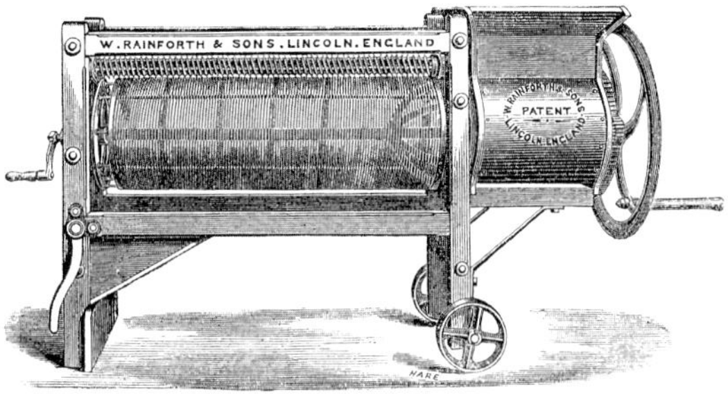

Fig 188
Rotary adjustable corn-screen, Penney & Co, Lincoln, 1878

Fig 189
Patent parallel lifting cultivator, W Rainforth & Sons, Lincoln, 1879

Fig 190
Potato digger, Penney & Co, Lincoln, 1879

BOOKS

This short list of books offers a general background to farm machinery and covers the Lincolnshire context. Also included are the few books that have been published about individual Lincolnshire firms. (Significant deposits of company records are accessible at Lincolnshire Archives, and some other printed material, as well as surviving machinery, is held at the Museum of Lincolnshire Life.)

Brown, Jonathan
Farming in Lincolnshire 1850-1945
(Lincoln, Society for Lincolnshire History & Archaeology, 2005)

Clark, Ronald H
Steam Engine Builders of Lincolnshire
(Lincoln, Society for Lincolnshire History & Archaeology, 1998)

Fussell, G E
The Farmer's Tools: The History of British Farm Implements, Tools and Machinery 1500-1900
(London, Andrew Melrose, 1952)

Lane, Michael R
The Story of the Wellington Foundry: A History of William Foster & Company
(London, Unicorn Press, 1997)

Lane, Michael R
The Story of the Britannia Iron Works: William Marshall Sons & Co, Gainsborough
(London, Quiller Press, 1993)

Newman, Bernard
One Hundred Years of Good Company
(Lincoln, Ruston & Hornsby, 1957)

Partridge, Michael
Farm Tools through the Ages
(Reading, Osprey, 1973)

Pointer, Michael
Hornsby's of Grantham, 1815-1918
(Grantham, Bygone Grantham, 1978)

Pointer, Michael
Ruston & Hornsby: Grantham, 1918-1963
(Grantham, Bygone Grantham, 1977)

Wright, Neil R
Lincolnshire Towns and Industry, 1700-1914
(Lincoln, Society for Lincolnshire History & Archaeology, 1982)

INDEX

Page numbers in bold refer to illustrations

A
Abbott & Co (Newark, Notts) 88
acetylene light 115
AGE consortium 15, 18
agent 43, 73, 74, 79, 81, 87, 88, 113, 120, 125, 138
Albans, St, Duke of 91
Amos & Smith Ltd (Hull) 71
Ancholme Navigation 85, 87
Ancholme, river 91, 114
Ashby, Thomas Woodhouse (Stamford) 9-11
Ashton, William (Horncastle) 43
Australia 12, 131, 138, 140
Aveling & Porter (Rochester, Kent) 15
Avery (Birmingham) 81

B
baler **119**
Bamford, Joseph (Uttoxeter) 115
Barford & Perkins (Peterborough) 15
Beacock, Matthew (Winterton) 66, 68
beam engine 121, **136**
bells, church 99
Bentall, E H (Maldon, Essex) 81
Berridge, William (Brigg) 76
binder **58**, 63, 69, 70, 114, 117
Binnington, John (Hull) 112, 116
Birch, James (Lincoln) 103-104
blackheart casting 105
Blackstone, Edward (Stamford) 11-15
Boston: Centenary Methodist Church 99; Corporation 131; docks 141; gas works 96, 139; Grand Sluice 96, 101; Market Place 131; Shodfriars Hall 141; Town Bridge 96; Witham Iron Works 97, 100
Boydell, William, engine **137**, 139
Branston 89
bridges **85**, 87, 96, 133, 136
Brigg: The Angel 115; Bigby Street 85; cemetery 86; Grammar School 87; Market Place 115; Otter's Mill 81; Pingley PoW Camp 118; Railway Station 118; waterworks 81
British Empire Exhibition (1924) 17
Brown & May (Devizes, Wilts) 128
Burrell, Charles (Thetford, Norfolk) 123
Burton, Edward (Lincoln) 46, 48
Butlins 129
Buxton 91

C
Caistor 77, 82
cannon **99**
carrot grading machine 65
cart 25, **26**, 30, 33, 34, 41, 48, 113, 114
Carter, Frank and Tod (Billingshurst, Sussex) 13, 14, 17-19
chaff cutter 8, **9**, 10, 11, 62, 63, 64, 69, 70, **71**, 88
Clayton & Shuttleworth (Lincoln) 9, 21, 46, 47, 53, 67, 80, 95, 98, 103
clock, church 66
clod crusher 79, 134
Collett & Co (Boston) 141
combine harvester 67, 73, 74, 117, 118, **120**, 123, 124
Conservative Party 38
Cooke family: John (1821-87) 24, 25, 32; Frank (1861-1922) 32-34; Sidney 34; Horace 34
Cooke, John 62, 65
corn drying 74, 125
Corringham 119
Coultas family: James (1788-1862) 36, 37; James (1819-1890) 37-40; John 37; James Perkins (1845-1910) 38, 39, 43, 44; William 39; Benjamin 39, 40; John Vickerman 39, 40; James Riley (1880-1952) 39
Coultas, James 124, 127
Coupland, Fred (Carrington) 127
Cousins (Emneth, Norfolk) 127
Crimean War 78
Crowley, John (Sheffield) 88
cultivator 33, 73, 114, **153**

D
Dairy Show 50
Day & Osgerby (Brigg) 88
Deering 114
Depression, Agricultural (1880s) 70, 141; (1920s-1930s) 71
Dossor & Weddall (Grimsby) 87
drainage 128, 136, 140
dray 25, **26**, 33, 34
dresser, corn **13**, 43, 79, 80
drills 8, **41**-44, **45**, 48, 63, 64, 65, 68, 70, 71, 79, 87, **152**
dryer corn 74
Duckering family: Richard (1814-70) 46-49; Charles (1841-1916) 48-53; Richard (1878-1964) 52-53

E
Eagle 24
Easton, Amos & Anderson (Reading, Berks) 89
Edlington family: John Butler (1839-1921) 56-61; Thomas (b1832) 56-59; Thomas E, 59-62; Thomas III, 61; Jack, 61; Alfred 62, Brian 62; Paul 62.
Edlington, J B (Gainsborough) 126
Edward VII, King 51
elevators 12, **13**, 20, 41, **43**, **60**, **102**, 123, **124**, 140
Emerson, James (Springfield, Mass., USA) 91
employee numbers 23, 25, 30, 33, 35, 37, 48, 50, 52, 54, 59, 66, 70, 74, 77, 82, 98, 100, 107, 108, 110, 126, 127, 131, 136, 140
employee wages 25, 70, 78
excavators 74
Exeter, Marquis of 11
export market 9, 29, 30, 33, 43, 64, 90, 92, 95, 108, 110, 129, 133, 134, 137, 138, 140

F
fairground rides 128, **129**, 130, 140
fire disaster 11, 37, 46, 60, 134-135
Fischer, Georg (Switzerland) 109-110
Fletcher family: John (1820-95) 66-69; Thomas (1850-1925) 69-71; John (1853-1907) 69; Henry John (1875-1935) 71-72; Thomas (1904-85) 72-73; John (1907-1991) 72-74; John Thomas (b1933) 74, Richard (b1961) 74
Ford Motor Company 110
Foster, William (Lincoln) 38, 53, 103, 121, 122, 123, 125, 126
Freeman, Percy (Lincoln) 53
Fuelol 17
Fulsby 87

G
GEC Alsthom 23
Gilkes, Gilbert (Keswick, Cumbs) 91, 93
Gill & Rundle (Tavistock, Devon) 121
Grant, Joseph Cooke (Stamford) 8
Grantham: Mowbeck 39; Museum 44; Perseverance Iron Works **37**, 39, 40; railway station 40; Union St. 36, 39; Wharf Rd, 37
Gratton, DT & Sons (Boston) 43
Great Exhibition 1851 9

H

Harrison, Frederick (Lincoln) 103-104, 106
Harrison, Thomas (Lincoln) 42
harrows 11, 25, 33, 43, 62, 63, 73, 79, **156**
Hart family: James (1791-1848) 76; William (1818-1898) 76-79, 82
Hart, James & Son (Brigg) 85, 87, 89
Hatfield (Yorks) 119
hay collector 63
haymaker 9, 11, **14**, 18, 20, 62
Hawker-Siddeley 21, 23
Head, Thomas (London) 87
Heckington 141
Herbert, Alfred (Coventry) 20
hermaphrodite **24**, 25
Hett, Charles Louis (Brigg) 79, 85-93, 113
Hick, Benjamin (Bolton, Lancs) 85, 86
hoist 125
Holbeach 37
Holmes & Son (Norwich) 42
Hornsby, Richard (Grantham) 9, 30, 37, 38, 39, 43, 45, 81, 88, 124, 134
horse gear (or works) 10, 11, **43**, 62, 63, 67, 68, 95, **115**
horse rake **8**, 10, **14**, 41, 56, 63, 64, 79, 114
horse hoe 8, 25, 33, 40, 43, 62, 63, 64, 68, **152**
hot bulb ignition 14, 16
Howard, J & F (of Bedford) 27, 30, 81
Howden family: William (1774-1860) 95-100, William (b1807) 100-101
Howden, William (Boston) 134
Hull Cart & Wagon Company 112
Hunt, R (Earls Colne, Essex) 126

I

incinerator 64
Institute of Civil Engineers 86
Institute of Mechanical Engineers 50, 53
irrigation 19, 22, 140

J

Jackson (Brigg) 86
JCB 119
Jeffery, George Edward (Stamford), 11

K

Kesteven County Council 38
kibbler 20
Kirton in Lindsey 90
kitchen ranges 39, 44, **48**, 49, 50, 51, 99

L

Laceby 119
lawnmower 79
Leys Malleable Castings Ltd (Derby) 105-109
Lincoln: Brayford Wharf East 104; Cattle Market 24, 50; City Council 52; Corn Exchange 50; Lindum Plough Works 24; Monks Road 24, 52; Railway Station 24; St Swithin parish 52; Waterside South/North 46, 48
Lincoln Castings 110
Lincolnshire Agricultural Society Show 30, 31, 41, 63, 64, 67, 69, 87, 88, 89, 90, 91, 113, 114, 115, **120**, 124, 125
Lincolnshire Rewinds (Lincoln) 130
Lister, RA, (Dursley, Glos) 18-21, 71
Louth 119
Luke, Robert (Stamford) 11
lurry 25, 34

M

MAN B&W 23
Manchester Ship Canal 92
manure distributor 41-44
marine engines 15, **18**, 19, 21, 98
Market Rasen 119
Marshall, William (Gainsborough) 56, 59, 68, 69
Massey-Harris 18, 21, 70, 73, 114, 118
Massey-Ferguson 70, 119, 120, 130
May, C A (Boston) 127
Maynard, Robert (Whittlesford, Cambs) 126
Meikle, Andrew (Dunbar, Scotland) 95
Methodist Church 36, 46, 52
mile post **99**
milk float 112, 114
mill, bone 10
mill, corn 10, 11, **13**, **20**, **23**, **49**, 51, 53, 62, 63, 64, 67, 68, 74
millwright 36, 72, 73, 74, 78, 82, 83, 85, 95, 98, 100, 101, 140
Mirrlees National Ltd (Stockport, Ches) 21
mower, grass **60**, 61, 63, 64, **68**, 70, 114, 115
munitions **15**
Museum of Lincolnshire Life 44

N

Naylor, John (Winterton) 66
Nelson, James & Sons (Leeds) 112
Newboult (Great Casterton) 12
Newsum (Gainsborough) 60
Normanby Hall Museum 66
North Lincolnshire Agricultural Society 80, 85, 87

O

oil cake crusher 11, 43, 63, 64
oil engine 12, **13**
oil mill 63

P

Paris Exhibition 10, 42, 136
patents 27, 38, 43, 56, 67, 68, 86, 89, 90, 93, 132, 133, 134, 136, 140
pea vining 64, 127
Peacock family: Henry EC (1866-1938) 112-117; Henry Allen (1899-1985) 117-119; Michael (b1933) 119-120
Peacock & Binnington (Hull & Brigg) 93
pearlitic malleable iron **108**
Penistan, Michael (Lincoln) 81
Penney & Porter (Lincoln) 104
petrol engine 18, 20
pig trough 33, **126**
pile driver **135**, 136
Pilkington, Thomas (Stamford) 8
Plano Manufacturing Co (USA) 114
plough **25-34**, 48, 49, **63**, 64, 113, 114, **115**, 117
ploughing match **28**, 30
Pocklington, John (Heckington) 141
pollution 10, HE2, 110, 138
portable steam engine **10**, **11**, 41, **43**, 67, 68, 78, **79**, 80, 81, 82, 91, 95, **98**, 100, 134, **135**, **136**, 140
Porter & Co. (Lincoln) 82
potato lifter 20, 21, 62, 64, **153**
potato planter 43
potato sorter 33, 61, 62, **64-65**
power generation 20, 21
PoWs 73, 118
press, furrow/land 64, 87, 127, **128**
pulveriser 14
pump, centrifugal 15-16, **89**
pump, water 81, **88-92**

R

railway, Spalding light agricultural 115
railway, locomotive 21, **22**, 53; signals 78, **80**, 81
Rainforth, William (Lincoln) 47
ram pump 81, 88-89, **91**
Ransomes, Sims & Jefferies (Ipswich, Suffolk) 27, 30, 69, 81, 114, 115, 117, 119
reaper, corn 41, 60, 62, **63**, 64, **68**, 114
Redbourne 74, 81, 91
reeing machine 132-133
Rennie, John 95-96
Richmond & Chandler (Manchester) 125
Robey, Robert (Lincoln) 103, 104, 126

Roller/rolls: flat 11, 14, 25, 33, 43, 49, 79; Cambridge 25, 33, **34**, 43, 62, 63, 65, 73, **74**, 81, 87, 126, 129
Roman villa (Winterton) 66
Roy, Rev William (Skirbeck, Boston) 132, 138
Royal Agricultural Society of England: implement and machinery trials 12, 24, 27, 41, 42, 80; shows 13, 30, 41, 42, **43**, 44, 51, 61, 63, 65, 80, 88, 90, 114, 140
Royal patronage 36, 40
Ruddock, J W 51
rulley 114
Rundle family: John Harness (d1974) 121-128; John H (1921-90) 123-128; Jack 128, 6, Ken 128; Sheila 128; Alan 128
Rundle, J H (New Bolingbroke) 39, 45, 99
Ruston, Proctor & Co (Lincoln) 39, 46, 47, 103
Ruston & Hornsby (Lincoln & Grantham) 39, 67, 117

S

sack barrow 113, 114
sack lifter 43, 44, 113, 114
Sainty, John (Burnham, Norfolk) 42
Samuelson (Banbury, Oxford) 81
Savage, Frederick (Kings Lynn, Norfolk) 130
saw bench 11, 140
scarifier 25
scoop wheel 91
Scotter windmill 87
seed drills 41-44, 79
Short, John (Boston) 100
showman's engine 122, 128
Skirbeck 131, 140
Sleaford Navigation 92
sluice works 136
Smith, Edward H (Brigg) 78, 87
Smith, Henry (Stamfords) 8-10
Smith & Grace Screw Boss Pulley Co (Thrapston, Northants) 10
Smithfield Show 10, 43, 44, 114, 115
social activities 10, 15, 20, 111
soil sifter 62
Spalding 133
Spight, Isaac (Brigg) 59, 71, 83
Spittlegate, Grantham 38
spring injection ignition 16, 18, 20
Sprowston (Norfolk) sugar beet machinery trials 61
Standen Engineering Ltd (Ely, Cambs) 73
Stamford: All Saints Iron Works 12; Borough Council 10; Broad Street 14; Museum 12, 20; Rutland Terrace Ironworks 9; Ryhall Road 11, 12, **19**; St Peter St. 8, **9**, 10; Sheepmarket 8
steam engine 63, 81, 82, 87, 88, 90, 95, 96, 98, 100, 140 (also see portable steam engine)
steam ploughing & cultivation 42, 82, 86, 117, 140
steam drill 42
steam sawing 121
steam tractor 121, 122
steam wagon 123
steel rolling mill 17
Stephenson, Thomas & Son (Boston) 123
street furniture **44**, **45**, 52, 53, **54**, 95, 99, 126, 140
sugar beet harvester 31, 61, 62, 64, 73
Sugden, W (Barking, London) 126
swath turner **14**, 79

T

tank 53
Taylor, Caleb (Boston) 135-136
Teague, John (Lincoln) 103-104
telephone 115
Thornycroft lorry 17
threshing machine 10, 36, 41, 48, 56, 66, 67, 68, 79-81, 95, 98, 99, 113, 123, 124, 125, 134, 135, 136, 138, 139, 140
threshing gang/contract work 66, 67, 68, 74, 123
Timberland 136
Tong, E J (Spilsby) 126
tractor 33, **34**, 73, **117**
traction engine 10, 14, 115, **122**, **133**, **137**-138, 140
trade mark **51**
Trent, river 113
trusser **50**
turbine, water 87, 90, **91**, **92**, **93**
turnip drill 42
turnip pulper/cutter 11, **21**, 43, **58**, 60, 61, 62, 63, 64, 71, **72**, 79, 88
Tuxford family: William Wedd (1782-1871) 131-133, 138, 140; James Edward (1774-1855) 131, 136; Wedd (1806-94) 133, 141; Joseph Shepherd (1808-78) 134, 136, 141; Weston (1814-1885) 133, 134, 141; William (1820-c1882) 134, 141; George Parker 140
Tuxford, William & Sons (Boston) 9, 79-80, 98, 99, 100

V

vaporiser 13, 14
vertical steam engine **12**, 14, 17, 20, 21, 63, 90

W

wagon 25, 30, 33, 34, 48, 113
Waterlow & Son (London) 81
watermill 79, 81, 87, 91, 92, **115**
waterwheel **87**, 88, 89, 92, 95, 140
waterworks 62, 81, 89
weighers 63
Welland, river 8, 15
whiteheart casting 104-105
Wicksteed Park, Kettering 130
Wilkinson, William (Boston) 100-101
Wilkinson, Wright & Co (Boston) 100-101
winnower 12, 43, 66, 68, 79
windmill 56, 67, 69, 79, 82, 87, 121, 131, 132, **133**, 134, 135, 136, 141
windpump **86**, 88
Wisbech 37
Winterton UDC 69
Witham, river 96, 97, 98, 134, 139
Women's Land Army 118
World War I 15, 20, 53, 60, 71, 116, 121
World War II 20, 53, 62, 65, 73, 111, 117
Wrangle Show 98
Wright, Henry (Boston) 100-102

Y

Yannedis & Co (London) 126
Yarborough, Earl of 76, 78, 79, 82, 113, 119
Yorkshire Show 41, 113, **114**

Z

Zoological Photographic Society 86

Fig 191
Zig-zag harrow, W Rainforth & Sons, Lincoln, 1883